W9-BYK-253

KOALAS
The little Australians we'd all hate to lose

KOALAS

The little Australians we'd all hate to lose

Bill Phillips

Australian National Parks
and Wildlife Service

An AGPS Press publication
Australian Government Publishing Service
Canberra

The National Library of Australia Cataloguing-in-Publication data:
Phillips, Bill.
Koalas

Bibliography.
Includes index.
ISBN 0 644 09697 7.

1. Koala.
2. Koala – Diseases – Australia.
3. Koala – Australia – Geographical distribution. I. Title.

599.2

Cover photograph: J. Sandon

Typeset in Australia by Canberra Publishing & Printing Co., Fyshwick, A.C.T.
Printed in Australia by Inprint Ltd.

FOREWORD

The Australian landscape is vastly different from when European settlers first set foot here just over 200 years ago. While the nation has prospered and become highly urbanised and technologically advanced it has been at the expense of large areas of forest and wildlife habitat. As a result over a hundred of our native plants and animals, including eighteen mammal species, are now considered extinct. The suggestion of just a few years ago that koalas may be headed down the same path attracted immediate attention from a broad cross–section of the Australian community and prompted the establishment of the Koala Conservation Program in 1985.

The Australian National Parks and Wildlife Service was pleased to accept responsibility for administering and coordinating the Koala Conservation Program which aimed primarily to examine the plight of koalas in modern Australia. The program gained initial support through the generous donation of $200 000 by American Express International. With additional input of public contributions and Commonwealth government funds, a comprehensive program of research and surveys was undertaken.

In this book Bill Phillips has brought together the most up-to-date information about koalas and presented it in a clear and concise way. He has also looked at the history of koalas: from how they came to be in Australia to their significance to the Aborigines, and from their 'discovery' by the Europeans to the impact of the koala fur trade during the 1800s and early 1900s.

Apart from describing the history of koalas and their fascinating private lives, this book presents a methodical and balanced appraisal of their current status based on recent research into the effects of the bacterium, *Chlamydia psittaci*, and the findings of the National Koala Survey which was undertaken as part of the Koala Conservation Program.

I recommend this book to everyone, not just people concerned for the future of koalas but everyone with an interest in wildlife conservation and the environment. I can only reinforce the observations and comments of the author, especially when he says that 'if current trends of land clearing are allowed to continue it will not only be koalas which suffer'. At a time when concern for our environment is growing daily, this book should serve as a timely reminder that the destiny of many of our native species will be determined by how mindful we are of their needs over the next decade.

J. D. Ovington

Professor J.D. Ovington
Director
Australian National Parks and
Wildlife Service

CONTENTS

Illustration Acknowledgments

Amble, A. and 'Greening Australia', 157
Atkinson, K., Auscape, 21
Australian Archives, Canberra, 13
Australian Information Service, Canberra, 3, 81, 82
Australian Museum, Sydney, 33, 34
Australian Wildlife Fund, 161
Australian Womens Hockey Association, 2
Ballarat Courier, 111
Beste, H. & J., Auscape, 142
British Museum (Natural History), 32
Burnett, D., Koala Preservation Society of Queensland Inc., 5, 65, 66, 67, 68, 69, 85, 86, 92, 94, 106, 115, 146, 150, 168, 173, 181, 182,
Bushing, V., Queensland Department of Environment and Conservation, 109
Cancalosi, J., Auscape, 132
Centre for Wildlife Research, courtesy S. Brown, 96, 97, 113, 136, 147, 154
Cheney, P. National Bush Fire Research Unit, CSIRO, Canberra,117
Clegg, J., Department of Anthropology, University of Sydney, 24
COO–EE Historical Library, Elwood, Victoria, 41, 42
Daily Mail, Brisbane, courtesy W.A. Pepperday Pty Ltd 52
Dick, T., Cambridge Museum of Archaeology and Anthropology, 31
Ellis, W., University of Queensland, 105
Featherdale Wildlife Park, Sydney, 4, 10, 172
Fell, P., 139, 141, 143
Ferrero, J–P., Auscape, 19, 20, 22, 53, 63, 93, 123, 144, 145, 183
Filmworld Research, Artarmon, NSW, 57, 58
Fox, A., 75
Gordon, G., Queensland Department of Environment and Conservation, 119, 120, 121
Gorrie, I., and 'Greening Australia', 152
'Greening Australia', 159, 170, 180
Haigh, A., 16
Handasyde, K., 61, 90, 91, 128, 155
Hutchings, P.T., National Bush Fire Research Unit, CSIRO,118
John Oxley Library, Brisbane, 14, 44, 45, 46, 48, 51
Jensen, M., Auscape, 137, 138, 148
Koala Preservation Society of New South Wales, 171, 175
Lewis, F., courtesy R. Warrenke, Arthur Rylah Institute, 73, 74
Lewis, G., Photo Library, Sydney, 35, 36, 59, 64
Lochman, J., 88, 116
Lone Pine Koala Sanctuary, Brisbane, 127, 130
Lumsden, L.F., Arthur Rylah Institute, Melbourne, 100
Martin, R., 60, 71, 124, 131, 140,
McGreevy, D.G., Queensland Department of Environment and Conservation, 149
Macmillan Publishing Co., The, 8, 9, 11
Mitchell Library (State Library of New South Wales), 43, 83, 84
National Library of Australia, Canberra, 15, 37, 38, 39, 40, 46, 47, 49, 50, 54, 55, 56, 78, 79, 87, 98, 99
Needham, W., 23
New South Wales Department of Agriculture and Fisheries, 160
New South Wales Department of Lands, 174
Obendorf, D.L., Tasmanian Department of Agriculture, 133, 134, 135
Parer, D. & Parer–Cook, E., Auscape, 1, 129
Parish, S., courtesy Zoological Parks Board of New South Wales, 158
Patterson, R., Queensland Department of Environment and Conservation, 89
Phillips, S., courtesy Koala Preservation Society of New South Wales, 156, 169
Phillips, W., Australian National Parks and Wildlife Service, 108
Prevett, P., Ballarat College of Advanced Education, 163
Queensland Newspapers Pty Ltd., courtesy *Courier Mail*, 7
Roberts, R.A., Art Australia, 30
Robertson, G., Auscape, 122
Retusa Pty Ltd., Angus and Robertson Publishers, 80
Sandon, J., 114
Schouten, P., 12, 18, 62, 95, 126
Smith, P., Warringah Shire Council, 165, 166
South Australian National Parks and Wildlife Service, 70
Starr, J., Koala Preservation Society of New South Wales, 177
Sun, Melbourne, 112
Totterdell, C., courtesy National Tree Program, 101, 102, 103, 162, 164
Trezise, P., 27, 28, 29
University of Queensland Press, courtesy G. Walsh and J. Morrison, 25, 26
Turner, L., Koala Preservation Society of New South Wales, 179
Victorian Naturalist, Field Naturalists Club of Victoria, 72
Whitehouse, D., Koala Preservation Society of New South Wales, 110, 176, 178
Wild Life, Herald and Weekly Times, 125
Wildlife Photography, courtesy G.B. Baker, 104, 107, 151, 153, 167,
Zoological Parks Board of New South Wales, 76, 77

ACKNOWLEDGMENTS

This book was possible because of the cooperation and involvement of many people and organisations. American Express International provided the financial support which gave the initial impetus to the Koala Conservation Program (KCP). Officers of the respective State and Territory wildlife and land management authorities assisted with and helped coordinate the National Koala Survey (NKS). The survey coordinators in each State deserve special commendation: Rodney Spark in South Australia; Christine Porter in Victoria; Peter Rubacek in New South Wales and Ross Patterson in Queensland.

Within the Australian National Parks and Wildlife Service (ANPWS), David Carter was the initial convenor of the KCP, followed by Frances Michaelis during the time when the NKS was underway. Lorna Keevers, Lynette Webb and Michael Crotty gave short–term assistance during the NKS. Special acknowledgment should also go to the thousands of volunteers who participated in the survey and made it a success.

Analysis of the results from the NKS was greatly assisted by the involvement and interest of Gunter Schmid and Ken Richardson (ANPWS, Computer Services Section). Richard Thackway (ANPWS, Scientific Programs Unit), Bob Kennard and Frank Bullen (Australian Surveying and Land Information Group) provided assistance with plotting koala distribution information and past and present tree cover. Peter Menkhorst and Ross Patterson were especially helpful with the survey data for Victoria and Queensland, respectively. Peter Tieman (ANPWS, Publications Section) advised on map design and specifications and all maps were prepared by Cangraphics, Canberra.

Thanks also go to the following people for making unpublished information available: Don Burnett (Koala Preservation Society of Queensland Inc.); Connie Adams (Avalon Preservation Trust); Peter Smith (Environmental Officer, Warringah Shire Council); Greg Gordon (Queensland Department of Environment and Conservation); Jean Starr (Koala Preservation Society of New South Wales); Wayne Murray (Bureau of Flora and Fauna); John Clegg (Department of Anthropology, University of Sydney); Lester Pahl and Ian Hume (University of Sydney); Bob Warneke, Gary Backhouse and Alan Crouch (Victorian Department of Conservation, Forests and Lands); Frank Carrick and Bill Ellis (University of Queensland); Keiran McNamara (Western Australian Department of Conservation and Land Management); Mark Lintermans (ACT Parks and Conservation Service); and Phillip Reed and Dan Lunney (New South Wales National Parks and Wildlife Service).

A special acknowledgment goes to Tony Lee, Roger Martin and Kath Handasyde for allowing me to refer extensively to their unpublished findings of research into *Chlamydia* which was partially funded by ANPWS under a consultancy agreement. Steve Cork, Roger Martin, Kath Handasyde, Greg Gordon, Don Fletcher, Greg Fraser, Phillip Reed and Tony Robinson kindly read selected sections of the text and provided constructive comments.

I am grateful to Carol Cooper of the Australian Institute of Aboriginal Studies for her assistance with researching the section on koalas and the Aborigines. Thanks also go to Percy Trezise, Bill Needham and John Clegg for providing information and photographs to illustrate this section. Permission to reproduce the myth describing the arrival of the Thurrawal peoples in Australia was kindly granted by Reed Books Pty Ltd. Graham Walsh, John Morrison and the University of Queensland Press gave their permission for the story of Didane and accompanying artwork to be reproduced.

Film World Research, Artarmon, New South Wales, located early Movietone footage of koalas and provided suitable stills. Thanks also to Bob Warneke for searching the voluminous photograph collection of Fred Lewis for appropriate material. The Field Naturalists Club of Victoria provided an early edition of the *Victorian Naturalist* and granted permission for photographs to be copied from this. Allan Fox generously provided an early copy of *Tree Warden*. Patrick Cook is also gratefully acknowledged for preparing and donating several cartoons.

Finally, thanks go to Chris Mobbs (ANPWS, Interpretation Section) for his invaluable assistance, expertise and encouragement throughout this project. Helpful comments on the draft were provided by the Director of ANPWS, Professor J.D. Ovington and Tim Richmond (ANPWS, Assistant Director) gave constant support throughout the project. A very special thanks goes to Trish Phillips for her tolerance and comments on parts of the manuscript. Credit should also go to the typists who cheerfully laboured through many drafts: Maureen Carter, Leonie Ellis, Pat Protas and Olga Mangos. Staff of the Australian Government Publishing Service, and in particular Patricia Gibson and Jim Gray, are gratefully acknowledged for their cooperation and professionalism. And to the others I know I have overlooked, thank you.

Bill Phillips

INTRODUCTION

Few Australian animals can claim the universal appeal of the ones with the teddy bear looks — koalas. Their comical little faces have charmed the world and, to many, koalas are Australia's best-known ambassadors.

Koalas have a special place in the heart of all Australians. From Olympic dais, with koala mascots proudly displayed by triumphant athletes, to spellbound children fascinated by the adventures of Blinky Bill, these unique marsupials have become an integral part of Australian society.

The popular image of koalas has been used by private enterprise for decades with billboards and advertisements in glossy magazines frequently bearing their form. The growing Australian tourist industry also owes a debt to koalas. They sit with tolerant, bemused expressions, seemingly unconcerned by clamouring tourists with cameras clicking incessantly.

2

3

4

5

THE COMPLETE ADVENTURES OF

BLINKY BILL

The escapades of a mischievous little koala in the Australian bush.
Including *Blinky Bill, Blinky Bill Grows Up* and *Blinky Bill and Nutsy.*

WRITTEN AND ILLUSTRATED BY DOROTHY WALL

6

1. Are they really headed down the road to extinction?

2. The victorious Australian Women's Hockey Team at the 1988 Seoul Olympics chose the koala as one of its mascots.

3. Koalas are sometimes seen clinging to royalty as during the Duke and Duchess of York's visit to Australia in 1988.

4. Koala at Featherdale Wildlife Park — possibly Australia's most popular animal.

5. The Koala is the focus for Victoria Point, Queensland, Cub's parade.

7

6. Generations of Australian children have been fascinated by the adventures of Dorothy Wall's storybook character, Blinky Bill (From *The Complete Adventures of Blinky Bill*, written and illustrated by Dorothy Wall, Angus and Robertson Publishers, Sydney, 1939).

7. Pope John Paul II holds a koala during his visit to Australia in 1986.

12. In more modern times koalas have been characterised in many forms but none more cleverly than in Peter Schouten's movie star koala 'Eukalypta' (From *Koalas Bizarre* by P. Schouten, Angus and Robertson Publishers, Sydney, 1985).

8

10

11

EUKALYPTA

12

8. Norman Lindsay's koala character, Billy Blue Gum, was once a common figure in cartoons relating to national events such as the 1956 Melbourne Olympics.

9. Billy Blue Gum was also used to support the war effort.

10.The Koala proves its popularity at the Featherdale Wildlife Park, Sydney

11. Billy Blue Gum sporting top hat and tails.

BILLY BLUE-GUM TAKES TO THE GUN-TREE

9

But what is it that makes koalas so popular, so adored? Is it their cheeky, unblinking stare, their child-like embrace or simply their cute and cuddly look? Whatever it is, there's no disputing the special affinity we all feel with these remarkable animals.

Gazing down from swaying eucalypts, chewing nonchalantly on tender young leaves, koalas appear unaffected, even indifferent to the world

13 & 14. Koalas have always been popular in advertising.

15. Concern for the future of koalas is not new, as shown by this cartoon which appeared in the *Age* Thursday, 12 May 1938.

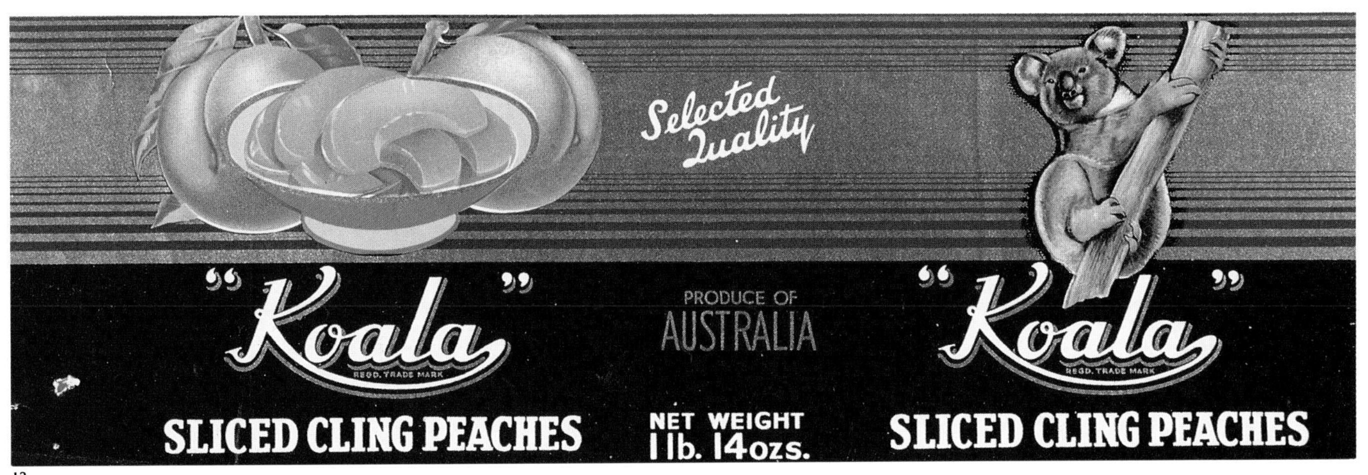

Selected Quality
"Koala"
REGD. TRADE MARK
SLICED CLING PEACHES
PRODUCE OF
AUSTRALIA
NET WEIGHT
1 lb. 14ozs.
"Koala"
REGD. TRADE MARK
SLICED CLING PEACHES

13

THE QUEENSLAND DIGGER. 1st November, 1928
South Brisbane
KOALA
(NATIVE BEAR)
Open Season Commences
Womens' Unanimous Decision
For Better Baking
USE
"KOALA" SELF RAISING FLOUR
An all Queensland Product
Printed at the office of H. POLE & CO. LIMITED; Elizabeth Street, Brisbane, for and on behalf of the R.S. and S.I.L.A. (Queensland Branch)

14

Youth to the Rescue.

15

around. However, as civilisation slowly closes in on their forest domain koalas can no longer take their peaceful existence for granted.

Before European settlement in Australia koalas were apparently abundant. The arrival of the British colonists heralded the beginning of an era of exploitation and harassment which saw koala populations exterminated in many areas. With intrusions into their gum tree world continuing, concern for the future of the species has grown. The last decade has seen an explosion of public interest in the welfare of koalas, largely fuelled by claims that the species is doomed due to the unrelenting removal of their vital forests and the lethal effects of the bacterium *Chlamydia psittaci*.

Of course, this latest outcry is not the first time that concern for the future of koalas has been voiced. In the mid 1800s, John Gould, the eminent naturalist, showed uncanny foresight when he expressed pessimism for the future of koalas because of their patchy distribution. Koala populations were ravaged by disease in the late 1800s and early 1900s, and the impact of hunters and the fur trade, leading up to the last 'open season' in 1927, is still evident in some parts of the country today. Koalas have shown remarkable resilience in recovering from these early depredations but there are signs that unless appropriate steps are taken now, they may soon become a rare sight in the Australian bush.

In response to growing public concern for koalas, officers of the Australian National Parks and Wildlife Service held discussions in late 1984 with experienced biologists, veterinarians and State wildlife authorities. This expert group recognised the seriousness of the threats to koalas but conceded that as so little was known about how they coped with life in the treetops and how they interacted with each other, it was impossible to recommend specific remedial action. The group therefore stressed the urgency of a major research effort directed at improving the general understanding of koalas so that suitable conservation measures could be devised.

In June 1985, the then Minister for Arts, Heritage and Environment, Mr Barry Cohen, launched the Koala Conservation Program. The Australian National Parks and Wildlife Service accepted responsibility for administering the program and a coordinated study program and national survey of koala populations was undertaken supported by Commonwealth funds, public contributions and a generous donation of $200 000 from American Express International. The research supported through the Koala Conservation Program aimed at identifying the specific habitat needs of koalas so that the management of populations in the wild could be improved. It was also recognised that there was a need to describe the characteristics and document the effects of the bacterium *Chlamydia psittaci* on koalas. Accordingly, financial support was provided to a range of research programs addressing a wide variety of subjects including koala food preferences, reproduction, forest needs, disease and disease transmission from koala to koala.

The National Koala Survey was undertaken during the 1985–86 and 1986–87 financial years. Support for conducting surveys of their koala populations was given to the conservation agencies in South Australia, Victoria, New South Wales, Queensland, Western Australia and the Australian Capital Territory. Under the Commonwealth Community Employment Program, temporary staff were employed and operated as State coordinators for the survey within the wildlife authorities of South Australia, Victoria, New South Wales, and Queensland. As testimony to the popularity of koalas, public support for the survey was overwhelming with between 10 000 and 20 000 people assisting in some way.

This book presents the findings of the Koala Conservation Program and National Koala Survey, as well as examining the history of koalas in Australia from earliest times through to the arrival of the first white settlers and the subsequent 200 years. The Australian continent has been vastly modified in this time and it is clear that all Australians have a role to play in ensuring that koalas remain a living part of our society for future generations to enjoy.

Minister for Environment
Parliament House
Canberra.

Paddington School
Oxford St.,
Paddington 2021
1/7/88

Dear Sir,

We in Class 3-H Wish to express our concern at the plight of the Koala To lose this animal would be tragic.

What can we do to help?

Your Sincerly
Alison Haigh.

16

16. In recent years community interest in koalas has reached new heights.

Help save an Australian we'd all hate to lose. Use the American Express Card.

17

17. The Koala Conservation Program, administered and coordinated by the Australian National Parks and Wildlife Service since 1985, gained initial funding support of $200 000 from American Express International.

KOALAS YESTERDAY

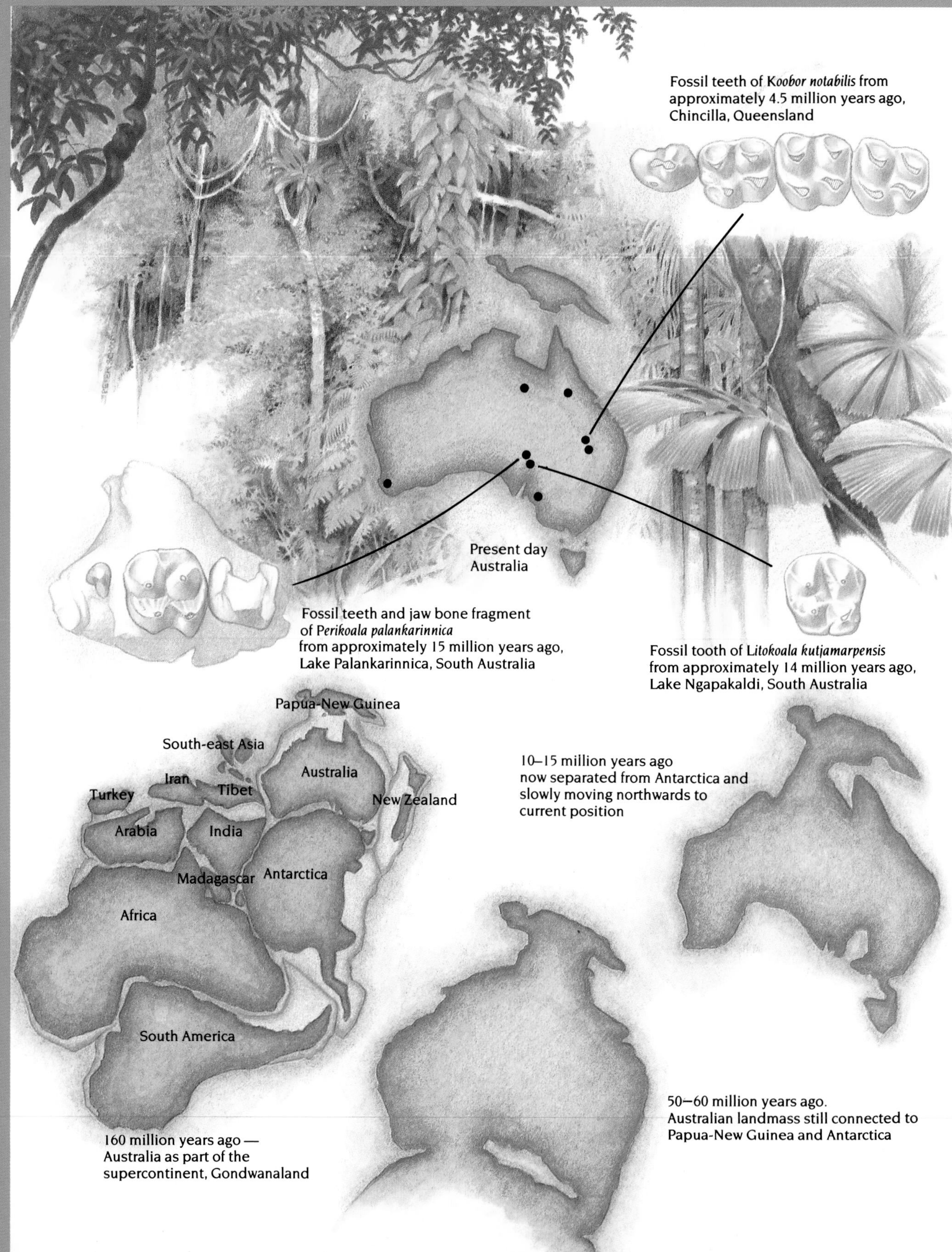
Fossil teeth of *Koobor notabilis* from
approximately 4.5 million years ago,
Chincilla, Queensland
Present day
Australia
Fossil teeth and jaw bone fragment
of *Perikoala palankarinnica*
from approximately 15 million years ago,
Lake Palankarinnica, South Australia
Fossil tooth of *Litokoala kutjamarpensis*
from approximately 14 million years ago,
Lake Ngapakaldi, South Australia
Papua-New Guinea
South-east Asia
Iran
Turkey
Tibet
Australia
New Zealand
Arabia
India
Madagascar
Antarctica
Africa
South America
10–15 million years ago
now separated from Antarctica and
slowly moving northwards to
current position
50–60 million years ago.
Australian landmass still connected to
Papua-New Guinea and Antarctica
160 million years ago —
Australia as part of the
supercontinent, Gondwanaland

HOW KOALAS CAME TO AUSTRALIA

Australia has a diverse fauna, with many species unique to our island continent. How Australia came to have these interesting and unusual species remains a contentious issue, although the Theory of Continental Drift possibly provides the best explanation.

According to this theory, 250 million years ago the world consisted of one landmass called Pangaea. In time, Pangaea broke up and one of the pieces, Gondwanaland, drifted into the southern hemisphere taking with it representatives of a number of animal groups. Gondwanaland in turn broke up and formed new landmasses, now recognised as Africa, India and New Zealand. The remaining Gondwanaland consisted of South America and Australia, as we now know them, joined by a landbridge destined to form Antarctica.

There were no polar icecaps at this time and parts of Gondwanaland were covered with rainforest. It is not known when or where the first marsupials evolved although it seems likely they came to Australia from South America via the then forested Antarctic landbridge.

About fifty-five million years ago Australia began to drift slowly northwards, gradually separating from the Antarctic landmass. The icecaps had still not formed, the seas were warm and Australia was largely covered with rainforest. The marsupials had little competition in this lush environment and new species unique to Australia evolved to occupy the many and varied ecological niches available. Koalas or koala-like animals probably first appeared during this period.

The polar icecaps began to develop about forty million years ago, around the time when Australia was finally separating from Antarctica. The seas gradually cooled, rainfall declined and Australia became drier. This cooling off period concluded twenty thousand years ago although fossil records indicate that significant climatic changes also took place about seven to eight million years ago.

Fossil records, from deposits in South Australia and Queensland, have, so far, revealed twelve different extinct koala species. Although only teeth and jaws have been found, it is clear that today's koalas are the only surviving members of a varied and large family of koala-like animals which once lived on the Australian continent.

Fossil material indicates that the early koalas, like those of today, were leaf-eaters, although their habitats were quite different. The earliest forms, from fourteen million years ago, apparently lived in rainforests. Fossils of koalas have also been found in geological material from four to five million years ago when the vegetation was open sclerophyll forest.

Precisely when koalas developed a liking for eucalypt leaves is not clear. Palaeobotanists contend that the genus *Eucalyptus* evolved in rainforest and Archer and Hand (1987) suggested that koalas may have adapted to eating eucalypt leaves when Australia was largely rainforest. As the rainforests receded with climatic changes, the eucalypts remained and the koalas stayed with them.

18. One explanation for how koalas came to Australia is provided by the Theory of Continental Drift. Fossils of koala-like animals have been found in a number of areas outside the current distribution of koalas (see upper map), including the now arid, central South Australia and north-west Queensland. Although only teeth and jaw fragments have been found it is clear that todays koalas are the only living members of a once diverse and large family of koala-like animals. As further testimony to their once widespread distribution in Australia, sub-fossils of koalas have also been found in south-west Western Australia. It is perhaps surprising that no koala fossils have, as yet, been found in Tasmania.

19. Koalas are marsupials just like the best-known members of the group, the kangaroos.

20 & 21. Like the egg-laying platypus and echidna marsupials were once thought of as 'primitive' mammals. Today they are considered a group which has simply adapted in a different, but no less efficient, way to cope with environmental conditions.

THE MARSUPIALS

Marsupials were once considered a primitive mammalian group and were placed, in the evolutionary sense, between the egg-laying monotremes such as the echidna and the platypus, and the placental mammals such as ourselves. This misconception has been laid to rest over the last fifteen years. Researchers have demonstrated conclusively that marsupials are no less efficient than the so-called 'advanced' placental mammals but have evolved differently in responses to environmental conditions.

The name marsupial comes from the pouch or marsupium in which the females carry, protect and suckle their young. The early European explorers to South America were fascinated by the marsupials they found. The Spanish explorer Vincente Pinzon was possibly the first to describe a pouch-bearing animal. In 1500 he returned from Brazil and presented a female opossum to King Ferdinand and Queen Isabella. Pinzon failed to recognise that the young in the pouch were attached to nipples, claiming instead that the young were suckled from elsewhere on their mother's body.

The first description of an Australian marsupial was provided by Francois Pelsaert in 1628. Pelsaert's ship *Batavia* was wrecked on Houtman's Abrolhos, off Western Australia, where he encountered the Tammar wallaby (commonly known as the quokka). Although Pelsaert's understanding of the function of marsupial pouches was more accurate than Pinzon's, he nevertheless misunderstood the basic reproductive physiology of the marsupials.

20

19

Below the belly the female carries a pouch, into which you may put your hand; inside this pouch are her nipples, and we have found that the young ones grow up in this pouch with the nipples in their mouths. We have seen young ones lying there, which were only the size of a bean, though at the same time perfectly proportioned, so that it seems certain that they grow there out of the nipples.

(Quoted from: 'Introduction to the Marsupials' by S. Cork in *Koala: Australia's Endearing Marsupial* L. Cronin (ed), Reed Books Pty. Ltd., Frenchs Forest, New South Wales, 1987)

Interest in the pouch-bearing animals was generated by these early accounts. However, not until after Captain James Cook's voyage to Australia in 1770 did the scientific community begin to theorise about the relationships of marsupials to other animals. The naturalists who accompanied Cook, Joseph Banks and Daniel Solander, described three marsupials. These were a

22

21

22. Wombats are thought to be the closest living relatives of koalas. Like koalas their pouch opens towards the rear and the tail is reduced to virtually nothing.

KOALAS AND THE ABORIGINES

native cat and a rabbit-like animal, which was either a bandicoot or kangaroo rat, found at Botany Bay and the, so-named, Captain Cook's Kangaroo in northern Queensland. Banks and Solander can be forgiven for not immediately appreciating the relationship between such a diverse trio of marsupials.

By the early 1800s the scientific community had come to appreciate the major differences in the reproductive patterns of the monotremes, marsupials and placentals. During the mid 1800s the Theory of Evolution provided a convenient explanation for these differences and the concept that the monotremes, marsupials and placentals represented a living evolutionary progression soon became established in scientific writings.

The perception of the marsupials as primitive mammals persisted until only a few years ago. Today, they are recognised as a highly adapted group which have evolved differently from other mammals for at least eighty million years.

The marsupials and placental mammals differ in a number of anatomical ways but the major distinguishing feature is the development of their young at birth. Marsupial young are born relatively early in their development and spend a lengthy period being nurtured in the pouch before becoming independent. In contrast, placental young are retained within the uterus for longer and emerge more fully developed.

It is a common misconception that all marsupials have a pouch. Some species have only a rudimentary pouch, others have none at all and still others form a pouch for the duration of the breeding season only.

Wombats are considered the closest marsupial relative of koalas. They share a host of anatomical characteristics unique to herbivorous marsupials. Most notable among these common features are the pouch which opens backwards, the granulated rather than ridged foot-pads most tree-dwelling animals have, and the greatly reduced tail.

Koalas feature as an important part of Aboriginal culture and are central to many Aboriginal myths and legends. Koalas are referred to in many Dreamtime accounts which describe the arrival of Aboriginal groups in Australia. One myth describes the arrival of the Thurrawal (or Thurrawai) people on the southern coast of New South Wales.

Long before there were men or animals in Australia, the only living things that had eyes to see the vast continent were flocks of migratory birds. When they returned to their homeland far to the east, they told the animals, which at that time had the form of men and women, of the unending plains, the tree covered mountains, the wide, long rivers, and the abundant vegetation of the great land over which they had flown. The reports created such excitement that the animals assembled from far and near to hold a Corroboree and discuss the matter. It was decided that, as the land appeared so much richer and more desirable than their own, they would all go and live there.

The big problem was how to reach the land of promise. Every animal had its own canoe, but they were frail craft, well

23. Aboriginal rock painting

ROCK CARVING OF A KOALA MOTIF, BURRAGURRA, NEW SOUTH WALES

This carving is one of a set of over 200 located on Burragurra, near the central coast of New South Wales. The site was one of a series of initiation places in that area and besides the koala, it contains tracks, figures of kangaroos, symbols and both human and spirit figures. While some of the story contained at the site is known, it is not understood what significance the koala carving has. Needham (1988) has suggested that it could have been a totemic figure.

23

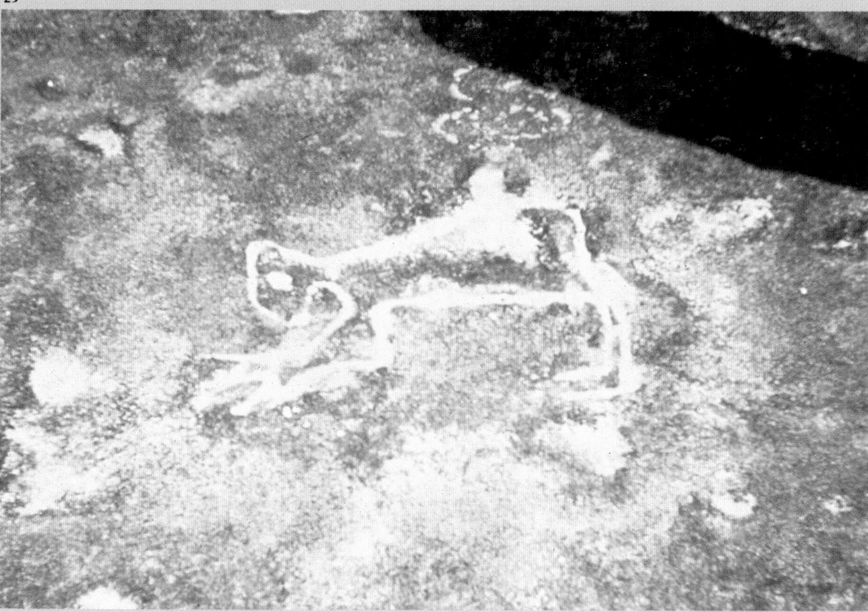

CULLAWINE, KARBOR, KOOLAH, COLO, KOALA?

The origin of the name koala is somewhat of a mystery. To the Aborigines koalas were known by a variety of names, largely as a result of regional differences in dialect. Among these names were cullawine, karbor, koolah, colo, colah, coolabun and koolewong. Other names showed no phonetic similarity to the word koala, for instance, boorabee, burroor, bangaroo, pucawan, banjorah and burrenbong (Stead 1934).

The early European settlers believed that koalas never drank water and because of this many considered that the word 'koala' was Aboriginal for 'no drink'. A number of Aboriginal legends describe the koala stealing water from the Aborigines. In some areas the settlers also believed that 'koala' meant 'biter' or 'angry' to the Aborigines.

It is possible that we owe the name 'koala' to an early typographical error although it seems more likely that the first European settlers adopted the name, or a phonetically similar one, from the Aborigines living near Sydney.

suited to the placid waters of lakes and streams, but not to the ocean that lay between the two lands. The only vessel that could contain them all was the one that belonged to Whale. He was asked if he would lend it to them, but he gave a flat refusal.

The animals were determined to migrate, no matter what difficulties had to be overcome. They held a secret meeting at which they enlisted the aid of Starfish, who was Whale's closest friend. Starfish agreed to help, for he was as anxious as the others to make the journey.

'Greetings, my friend,' he said to the Whale.

'Greetings,' Whale replied in his deep, rumbling voice. 'What do you want?'

'There is nothing I want, except to help you. I see your hair is badly infested with lice. I thought that as I am so small I could pick them off for you.'

'That's extraordinarily kind of you. They do worry me a bit,' Whale admitted. He placed his head in Starfish's lap and gave a sensuous wriggle of contentment. Starfish plucked off the lice in a leisurely fashion.

While the cleaning task went on, the animals went on tiptoe to the shore, loaded all their possessions in Whale's huge canoe, and paddled silently out to sea. The faint splash of their paddles was drowned by Starfish as he scratched vigorously at the vermin. After a while Whale became restless, and began to fret.

'Where is my canoe?' he asked. 'I can't see it.'

'It's here, right beside you,'said Starfish soothingly.

He picked up a piece of wood and struck a hollow log by his side. It gave out a booming noise.

'Are you satisfied now?'

ROCK ENGRAVINGS OF KOALAS — BEROWRA WATERS, NEW SOUTH WALES

Given that koalas feature in a number of Aboriginal myths and legends it is not surprising that koala-like forms or motifs also appear among the rock engravings and paintings of the Aborigines.

In and around Sydney there are many Aboriginal rock engravings which have been variously intrepreted as representing koalas.

The early work of Campbell (1899), from the Port Jackson, Brolson Bay region, describes a 'native bear' motif and the publications of McCarthy in 1956 and 1961 also identifies several rock engravings as 'koalas'.

More recently Clegg (1988) has documented a rock engraving from Berowra Waters, a northern suburb of Sydney, which he contends shows a mother and a young koala. Clegg is circumspect about his description, however and points out that

Aboriginal rock engravings are frustrating to look at . . . and frustrating to show people, because we do not know enough about them to answer the real and important questions that most people ask. We know they . . . were an important part of Aboriginal life at some time in the past. We guess that they were significant — holy might be a better word and we can only guess about any particular pictures's meaning or use.

(From: 'Berowra Waters Koala Engravings' in a paper presented to koala summit New South Wales National Parks and Wildlife Service, Sydney, November 1988.)

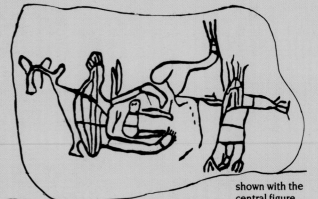

24. Aboriginal rock engravings at Berowra Waters near Sydney (left) and as redrawn by John Clegg (above). Various figures are shown with the central figure embracing a smaller form of similar shape possibly meant to represent a koala.

25. Didane the Koala, illustrated by J. Morrison.

Whale sank back again and submitted himself to his friend's attentions once more. The sun was low in the sky when Whale woke up for the second time.

'I am anxious about my canoe,' he said. 'Let me see it.'

He brushed Starfish aside and rolled over so that he could look round him. There was a long furrow in the sand where the canoe had been pulled down to the beach, but of the canoe itself there was no sign. Whale turned round in alarm and saw it on the distant horizon, almost lost to sight. He turned on Starfish and attacked him so fiercely that the poor fellow was nearly torn to pieces.

His limbs and torn flesh were tossed aside contemptuously. His descendants still hide among the rocks and salt water pools as their ancestor did that day, and their bodies bear the marks of the fury of the Whale when he turned against his friend. But little Starfish had not submitted to punishment without some resistance, and in his struggles he managed to tear a hole in Whale's head, which is also inherited by the descendants of their huge ancestor.

Whale raced across the ocean with water vapour roaring from the hole in his head, and began to overtake the canoe. The terrified animals dug their paddles deeper in the water and strained to make their canoe go faster, but it was mainly through the efforts of Koala that they managed to keep at a safe distance from their infuriated pursuer.

'Look at my strong arms,' cried Koala. 'Take your paddle strokes from me.' The gap grew wider as his powerful arms made the paddle fly through the water, and ever since his arms have been strong and muscular.

The chase continued for several days and nights, until at last land came in sight; the country they had longed for. At the entrance to Lake Illawarra the canoe was grounded and the animals jumped ashore. As they disappeared into the bush the canoe rose and fell on the waves.

Brolga, the Native Companion, was the only one who had the presence of mind to remember that they would never be safe while Whale was free to roam the seas in his canoe, for at any time he might come ashore and take up the pursuit again. So Brolga pushed the canoe out from the shore and danced and stamped on the thin bark until it was broken and sank beneath the waves. There it turned to stone; and it can still be seen as the island of Canman-gang near the entrance to Lake Illawarra.

Ever since that day Brolga has continued the dance that broke up the canoe.

Whale turned aside in disgust and swam away up the coast, as his descendants still do. As for the animal men, they explored the land and found it as good as the birds had said.

They settled there, making their homes in trees and caves, by rivers and lakes, in the bush, and on the endless plains of the interior.

(From: *Myths and Legends of Australia* by A.W. Reed, A.H and A.W. Reed, Sydney, Wellington, London, 1965.)

Some tribes greatly respected koalas and considered them wise counsellors from whom advice was sought on many matters. To the Bidgara people of the Carnarvon Gorge in central Queensland the koala called Didane was responsible for transforming their tribal lands from barren desert to lush, green forest.

Back in the Dreamtime, the rugged Carnarvon area was a very hot, dry place. There were no trees or bushes, and no grass.

When the first people arrived, the country seemed new and strange, with narrow gorges and the towering sandstone cliffs of Boodyadella, the main dividing range. The people came to love these craggy ranges, but were sad that no trees or grass grew.

Some animals were already living in the ranges — Ngaargoo the grey kangaroo, Waarunn the wallaby and Didane the koala. They, too, were sad about the dry treeless land.

The tribal elders met to discuss the problem. They wanted to bring trees and plants to this beautiful country. But how?

One wise elder suggested they try to get seeds from the trees growing in the sky. Perhaps a strong boomerang thrower could hit the trees and knock down the seeds.

The warriors of the tribe were called together, and the elders told them of their plan.

All the warriors wanted to help. Each thought he would be the one to knock down the seeds.

The whole tribe gathered round. One by one, the warriors moved to the centre of the group and threw their boomerangs as hard as they could.

As the people watched in silence, the boomerangs swirled upwards into the sky, but then fell back to earth.

After the last boomerang fell, the worried elders sat down again and talked about the problem. One wise old man with a white beard suggested they ask Didane the koala for help. With his broad chest and powerful arms the koala must be a good boomerang thrower.

Didane agreed to try. His friends Ngaargoo and Waarunn came with him to the place where the tribe had gathered.

Didane brought his largest war boomerang. Silence fell on the group as he prepared to throw it.

With a tremendous swing Didane hurled his huge boomerang up into the sky. Its swishing sound faded away as it passed through the clouds and out of sight. All

25

26. Didane the Koala, illustrated by J. Morrison

27. Noted expert on Aboriginal rock paintings, Percy Trezise sits against the backdrop of the desolate Mosman Gorge in northern Queensland.

26

eyes were fixed on the sky as they waited for the boomerang to return.

They waited a long time. The boomerang seemed lost forever. Some of the women began to weep. They knew that if Didane's powerful boomerang could not reach the trees, there would be no hope for their land.

Suddenly a shower of seeds began to fall. Seeds of every kind, large and small, rained down on the hot, dry earth.

With shouts of joy the people began to dance around Didane. He was now a hero.

Soon rain came, cooling the land and filling the rivers. The seeds knocked from the sky by Didane's boomerang began to grow in the fertile soil . . .

(From: *Didane the Koala* by G.L. Walsh, University of Queensland Press, St.Lucia, Queensland, 1985.)

KOALAS IN ABORIGINAL ROCK PAINTINGS IN NORTHERN QUEENSLAND

Although no longer found in the desolate interior of Cape York, the Aboriginal rock paintings of koalas found in the Mosman Gorge and the Laura River area suggests that they once lived there. Percy Trezise, noted expert on Aboriginal rock art, suggests that:

koalas may have disappeared from this area around 15 000 years ago during prolonged periods of drought which also saw the demise of the platypus in the Laura River. It is also possible koalas may have re-colonised the formerly drought-stricken regions of Cape York around five to eight thousand years ago when, during a more humid phase, rainforest was re-established on the Atherton Tableland and the area was recolonised by people.

(From: Personal Communication, 1988)

27

28

29

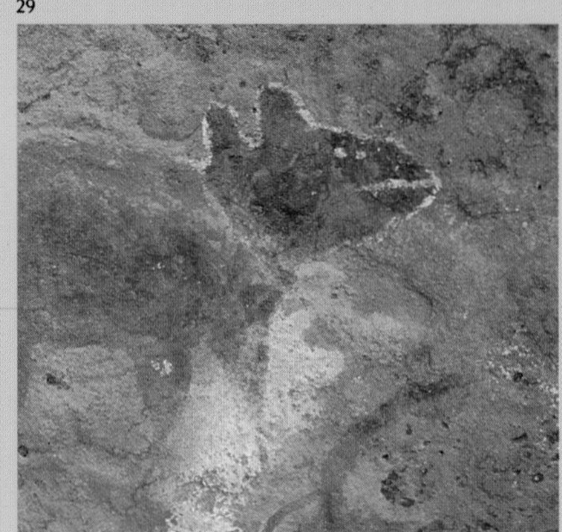

28. Although koalas are not found at Mosman Gorge today, Percy Trezise has found koala motifs amongst the Aboriginal paintings of the region.

29. An enlargement of one of the Mosman Gorge koala motifs.

30. From the painting *The Drought Maker* by Ainslie Roberts, 1964

30

A number of other legends describe the story of koalas stealing the precious water supplies from Aborigines. In *The Aborigines of Victoria*, R. Brough Smyth relates the story of the koala, Koob-borr, as told to him by the Aborigines of the Upper Yarra.

The tribe . . . were not kind to him. At one time water was very scarce everywhere, and poor little Koob-borr could not get any. No person would give him any water. On a certain day all the tribe went out to hunt, and they forgot to take little Koob-borr with them. . . . The people had forgotten to hang up their tarnuks — they were full of water — and for once Koob-borr had more than enough to drink. But that he might have always plenty, and also avenge the wrongs which had been done to him, he took all the tarnuks and hung them up on the boughs of a little tree. Having done this, he next brought all the water of the creek and put it into the tarnuks, and finally he climbed the tree and seated himself beside the tarnuks. The tree suddenly became very large . . . and Koob-borr sat in the tree until evening; and evening brought back the blacks. The blacks were very thirsty . . . they found that all the tarnuks had been taken . . . the creek was dry. Presently one of the men saw the big tree . . . and they saw their tarnuks hanging on the high boughs and little Koob-borr sitting in the midst of them.

'Wah!' says one, 'is that you? Have you any water there?' 'Yes,' replied Koob-borr, 'here am I, and I have plenty of water; but I will not give you one drop, because you would not give me any when I was nearly dying for the want of water.' They were all very thirsty and two of the men at length commenced to climb the big tree. Koob-borr laughed at them, and let fall a little water on them, and they loosened their hold of the tree and fell to the ground and were killed. Two men again attempted to climb to the bough on which Koob-borr was seated, but he treated them in the same way, and they too fell down and were killed. Two more attempted to climb, and again they fell down and were killed, and two more, until nearly all the men of the tribe were killed. Then men of other tribes came, and two by two they attempted to ascend, and Koob-borr spilled water on them, and they fell down and were killed. At length Ta-jerr and Tarrn-nin came to the relief of the blacks. They climbed round and round, just in the line which a creeping plant takes. Koob-borr laughed at the others, until they had ascended to a great height, and then he took water and let if fall, but the men were no longer in the same place, but higher up, and it did not fall on them. Koob-borr ran and got more water, and poured it where he had last seen the men, but again it did not touch them; and finally Ta-jerr and Tarrn-nin reached the high boughs. Koob-borr now began to cry, but they heeded not his cries. They seized him and beat him until all his bones were quite soft. They then threw him down, and other blacks beat and tried to kill him. He did not die. He became in form and appearance what he is now, and he ran up another tree. Ta-jerr and Tarrn-nin cut down the big tree in which the tarnuks and all the water were; and the water came out of the tree, and flowed into the creek and there has been ever since plenty of water.

From this time Koob-borr became food for the people; but it is a law amongst the people that they must not break his bones when they kill him, neither take off his skin before they roast him. If the law were broken, Koob-borr would again become powerful, and he would dry up the waters of the creeks.

(From: *The Aborigines of Victoria* by R. Brough Smyth, vol.1. John Currey, O'Neil, Melbourne, 1876.)

Despite the significant role of koalas in Aboriginal myths and legends they were a readily available source of food and one relished by Aborigines. Fountain commented that:

The young koala is much esteemed as an article of food by the blacks, who climb the trees and knock mother and young . . . from the branches. Though they often fall more than a hundred feet to the ground, they are never killed outright, and sometimes not even disabled. It is left to the knives and hatchets of the gins to complete the cruel tragedy.

(From: *Rambles of an Australian Naturalist* by Paul Fountain, John Murray, London 1907.)

Prior to the arrival of European settlers, there were apparently few koalas in areas supporting large Aboriginal populations. It is thought that koalas became more abundant in the forests where the first Europeans settled, as in these areas there were fewer Aborigines and predators such as dingoes. Nevertheless, there seems little doubt that, in general, koalas were abundant over their range prior to the arrival of the First Fleet.

31. In some areas koalas were a regular part of the diet of the Aborigines. This Aboriginal hunter is heavily laden with animals including a koala (hanging from his left shoulder). The photograph was taken by Thomas Dick in the Port Macquarie area in about 1915.

31

DISCOVERY OF KOALAS BY EARLY SETTLERS

From the available records it appears that when Captain Cook landed on Australian soil in 1770 he was not greeted by the sleepy inquisitive stares of koalas. This may be an oversight attributable to the fleeting nature of Cook's visit but it is more likely that koalas were scarce in the forests surrounding Botany Bay. It wasn't until ten years after the arrival of the first white colonists in 1788 that koalas were discovered.

On an expedition into the Blue Mountains in 1798, John Price, a free-servant of the Governor, Captain John Hunter, wrote of an 'animal which the natives call a Cullawine which much resembles the Sloths in America'. This first recorded encounter with a koala apparently took place near the present-day town of Bargo, ninety kilometres south-west of Sydney. Troughton (1967) reports that the date was appropriately 26 January — Australia Day.

Four years later, in 1802, a further koala sighting was recorded by Ensign F. Barrallier in his journal describing explorations into the inland of New South Wales. The report was not published until almost a hundred years later, as an appendix to the *Historical Records of New South Wales* (vol. v). In his report Barrallier stated:

> Gory told me that they had brought portions of a monkey (in the native language Colo), but they had cut it in pieces, and the head, which I should have liked to secure, had disappeared. I could only get two feet through an exchange which Gory made for two spears and one tomahawk. I sent these two feet to the Governor preserved in a bottle of brandy.'

(From: *The Early History of the Koala* by T. Iredale and G.P. Whitley, Victorian Naturalist, vol.51, Melbourne, 1934.)

In the following year Barrallier was successful in obtaining a live koala with two back-young, which he gave to Governor King for transportation to England. Apparently the koalas did not survive long in captivity and failed to reach their intended destination alive. The *Sydney Gazette* of August 1803 carried the following account of Barrallier's find:

THE SYDNEY GAZETTE, And New South Wales Advertiser.

PUBLISHED BY AUTHORITY.

Vol. I. SUNDAY, AUGUST 21, 1803. Number 25.

> An animal whose species was never before found in the Colony, is in His Excellency's possession. When taken it had two Pups, one of which died a few days since. This creature is somewhat larger than the Waumbat [sic], and, although it might at first appearance be thought much to resemble it, nevertheless differs from that animal . . . the graveness of the visage . . . would seem to indicate a more than ordinary portion of animal sagacity, and the teeth re-semble those of a rabbit. The surviving Pup generally clings to the back of the mother, or is caressed with a serenity that appears peculiarly characteristic: it has a false belly, like the apposum [sic], and its food consists solely of gum leaves, in the choice of which it is excessively nice.

It is interesting to note these first references to the koalas 'false belly' (pouch) as well as recognition of their fastidious dietary habits.

The first major account of koalas came from Everard Home and was

DISCOVERY OF THE KOALA

32. One of the earliest drawings of a koala was provided by Ferdinand Bauer, a member of a scientific party which accompanied Matthew Flinders on his voyage to Australia in 1801.

33. The rare book Perry's *Arcana* of 1810 contained an early account and sketch of a koala, then called 'Koalo' or 'New Holland Sloth'.

published in the *Philosophical Transactions* of 1808. Despite some obvious inaccuracies Home's remarks provide an insight into the early understanding of koalas:

The Koala is another species of the Wombat, which partakes of its peculiarities. The following account of it was sent to me some years ago by Lieutenant-Colonel Paterson, Lieutenant-Governor of New South Wales. The natives call it the Koala Wombat; it inhabits the forests of New Holland, about 50 or 60 miles to the south-west of Port Jackson, and was first brought to Port Jackson in August 1803. It is commonly about 2 feet long and one high, in the girth about one foot and a half; it is covered with fine soft fur, lead-coloured on the back, and white on the belly. The ears are short, erect, and pointed; the eyes generally ruminating, sometimes fiery and menacing; it bears no small resemblance to the bear in the fore part of its body; it has no tail; its posture for the most part is sitting.

The Koala feeds upon the tender shoots of the blue gum tree, being more particularly fond of this than of any other food; it rests during the day on the tops of these trees, feeding at its ease, or sleeping. In the night it descends and prowls about, scratching up the ground in search of some particular roots; it seems to creep rather than walk. When incensed or hungry, it utters a long, shrill yell, and assumes a fierce and menacing look. They are found in pairs, and the young is carried by the mother on its shoulders. This animal appears soon to form an attachment to the person who feeds it.

(From: 'An Account of Some Peculiarities in the Anatomical Structure of the Wombat,' *Philosophical Transactions of the Royal Society of London*, 1808.)

32

33

One of the earliest drawings of a koala was provided by Ferdinand Bauer. A gifted botanical artist, Bauer was chosen by Sir Joseph Banks to join a small contingent of scientists on the voyage of Matthew Flinders in 1801. Their aim was to survey and chart the coasts of Australia. When he returned to England in 1805 Bauer produced over 250 water-colours of Australian plants and animals, including the koala.

Another of the early accounts of a koala, which was accompanied by a sketch, appeared in 1810 in the rare book Perry's *Arcana*. Perry referred to the koala as the 'Koalo or New Holland Sloth' and likened koalas to the sloths, largely on the basis of their sleepy disposition and economy of movement.

Although it does not agree entirely, in the form of its feet, with either the three-toed or two-toed Bradypus (sloth) which are found in other countries, yet the similitude is so strong in most peculiarities, which it possesses, that the naturalist may perhaps be considered as fully justified in placing it with the Genus Bradypus or Sloth. It is necessary to repeat, that this animal, of which there are but three or four species known, has received its name from the sluggishness and inactivity of its character, and for its remaining for a long time fixed to one spot. It inhabits woody situations, where it resides among the branches of trees, feeding upon the leaves and fruit, and is a solitary animal rarely to be met with. It is armed with hooked claws and the fore feet are in general longer than the hinder ones: some of the species of Bradypus have a tail; others are without.

Among the numerous and curious tribes of animals, which the hitherto almost undiscovered regions of New Holland have opened to our view the creature which we are now about to describe stands singularly pre-eminent. Whether we consider the uncouth and remarkable form of its body, which is particularly awkward and unwieldy, or its strange physiognomy and manner of living, we are at a loss to imagine for what particular scale of usefulness or happiness such an animal could by the great Author of Nature be destined. That the solitary and desert wastes of that immense country should be animated by creatures of so different a texture and appearance to any hitherto known, no Naturalist, however sanguine in his expectations, could have easily suspected.

The Koala is supposed to live chiefly upon berries and fruits, and like all animals not carnivorous, to be of a quiet and peaceful disposition . . . The Koala has more analogy to the Sloth-tribe than any other animal that has hitherto been found in New Holland, the eye is placed like that of the Sloth, very close to the mouth and nose, which gives it a clumsy awkward appearance and void of elegance in the combination. The motions of such a creature being slow and languid, and the back lengthened out by the continual hanging posture which they assume; they have little either in their character or appearance to interest the Naturalist or Philosopher. As Nature, however, provides nothing in vain, we may suppose that even these torpid, senseless creatures are wisely intended to fill up one of the great links of the chain of animated nature, and to show forth the extensive variety of the created beings which GOD has, in his wisdom constructed.

(From: *Arcana* by George Perry, 1810.)

The next drawing of a koala was published in A *Companion to the London Museum and Pantherion* (17th Edition) in 1814 and was based on specimens held in the collection of William Bullock of Liverpool. Bullock assembled a private collection of Australian natural history specimens obtained from expeditions. The drawing shows a koala with an exaggerated back and rear.

A further illustration of a koala appeared in Cuvier's *Règne Animal* in 1817 and showed the koala in an unusual quadrupedal gait, with simplified facial features reminiscent of a cat. While it is amusing to compare these first representations of koalas to the animal with which we are all so familiar, it should be remembered that these early drawings were based on museum specimens and written accounts and not live animals.

34

34. This somewhat unusual representation of a koala appeared in Cuvier's *Regne Animal* in 1817.

SCIENTIFIC INTEREST IN KOALAS

Scientific interest in koalas was considerable following the publication of a sketch in Perry's *Arcana* in 1810. However, it was not until 1814, following a visit to London by French naturalist H.M. de Blainville, that koalas received the generic name of *Phascolarctos* from the Greek *phaskolas* for leather bag (pouch) and *arktos* for bear. The following year the German naturalist Goldfuss provided the specific name *cinereus*, from the Latin for ash-coloured, thus establishing the present scientific name.

The specimen examined by Goldfuss was from west of the Nepean River, near Sydney (Troughton 1941) and had the typical ashen-grey fur of koalas from that area. In 1923, Oldfield Thomas of the British Museum described a northern race of koalas (P. *cinereus adustus*) from south-east Queensland and Troughton in 1935 separated a further sub-species (P. *cinereus victor*) from Victoria. Today regional differences in fur thickness and colour are recognised. These sub-species or races are considered to represent the extremes of a north-south trend in body size with southern koalas sometimes weighing in at almost twice the weight of their northern counterparts.

35

36

35 & 36. There are regional differences in the appearance and size of koalas. Some Victorian koalas are twice the body weight of their northern counterparts. Those from the south tend to have darker and thicker fur.

37. The *Saturday Magazine*, 31 December 1836

THE
Saturday Magazine.

Nº 288. DECEMBER 31ST, 1836. PRICE ONE PENNY.

COMMITTEE OF GENERAL LITERATURE & EDUCATION

UNDER THE DIRECTION OF THE COMMITTEE OF GENERAL LITERATURE AND EDUCATION, APPOINTED BY THE SOCIETY FOR PROMOTING CHRISTIAN KNOWLEDGE.

SKETCHES OF NEW SOUTH WALES. No. XIV.

ON THE ANIMALS CALLED "MONKEYS," IN NEW SOUTH WALES.

THESE animals are common in New South Wales, and the accompanying sketch is a correct representation of one of them. They are generally found in thick stringy-bark forests, and are numerous on the ranges leading to Cox's River, below the mountain precipices, and also in the ravines which open into the Hawkesbury River, as well as in various other parts of the colony. They are called by some monkeys, by others bears, but they by no means answer to either species. I first took them to be a species of the sloth of Buffon, and so they might be, though they differ also in many respects from that animal; and I now think that these animals mostly resemble, and come nearest to, the loris, or slow-paced lemur of India.

Having shot several, and caught them occasionally (with the assistance of the natives) alive, both young and old, which I have kept at the tents for some time, I am able, from what I have observed, to give the following description. They have four hands, having naked palms, which are armed with crooked pointed nails, exceedingly sharp, and rather long. They are covered with fur of a bluish-gray colour, very thick, and extremely soft. It is darker on the back, and paler under the throat and belly, but slightly tinged with a reddish-brown about the rump.

VOL. IX. 288

37

A further early report of koalas came from noted botanist Allan Cunningham. In his journal for 9 November 1818, when exploring in the Illawarra region of New South Wales, he made reference to opossums which with hindsight were clearly koalas.

The native, our guide, espied, on a tree, an opossum (Didelphis), having many of the habits of the ring-tailed species (caudivolva). It was a female and her cub. They were asleep, hanging by the claws, among the topmost shoots of a slender *Eucalyptus piperita*. It has no tail; it has the thick bluff head of the wombat, with strong incisor teeth, but does not burrow in the earth as that harmless, easily domesticated animal. The length of the mother was 28 inches, and its weight upwards of 30 lbs; the cub was about half grown, its length not exceeding a foot; it was covered with a fine thick grey fur. The Australian killed the parent in order the better to carry her down the range, but the young one, at my suggestion, and request, was suffered to live, and was carefully brought to the Farm hut.

(From: *The Early History of the Koala* by T. Ireland & G.P. Whitley, Victorian Naturalist, vol. 51, Melbourne 1951.)

In the *Saturday Magazine* of 31 December 1836, William Govatt, an assistant surveyor to Major T.L. Mitchell, related some of his observations of koalas as part of a series entitled 'Sketches of New South Wales'. Govatt referred to koalas as 'monkeys', likening them to the loris lemurs of India.

ON THE ANIMALS CALLED 'MONKEYS' IN NEW SOUTH WALES

These animals are common in New South Wales, and the accompanying sketch is a correct representation of one of them. They are generally found in thick stringy-bark forests, and are numerous on the ranges leading to Cox's River, below the mountain precipices, and also in the ravines which open into the Hawkesbury River, as well as in various other parts of the colony. They are called by some monkeys, by others bears, but they by no means answer to either species. I first took them to be a species of the Sloth of Buffon, and so they might be, though they differ also in many respects from that animal; and I now think that these animals mostly resemble, and come nearest to the loris, or slow-paced lemur of India.

Having shot several, and caught them occasionally (with the assistance of the natives) alive, both young and old, which I have kept at the tents for some time. I am able, from what I have observed, to give the following description. They have four hands, having naked palms, which are armed with crooked pointed nails, exceedingly sharp, and rather long. They are covered with fur of a bluish-grey colour, very thick, and extremely soft. It is darker of the back, and pales under the throat and belly, but slightly tinged with a reddish-brown about the rump. The nose is somewhat elongated, and appears as if it was tipped with black leather. The ears are almost concealed in the thickness of the fur, but have inwardly long whitish hairs. The eyes are round and dark, sometimes expressive and interesting. The mouth is small, and they

38, 39 & 40. These illustrations and this account of the koala by noted naturalist John Gould in 1863 are among the most accurate of the 19th Century.

PHASCOLARCTOS CINEREUS.
J. Gould and H.C. Richter del. et lith.
Hullmandel & Walton Imp.

38

PHASCOLARCTOS CINEREUS.

39

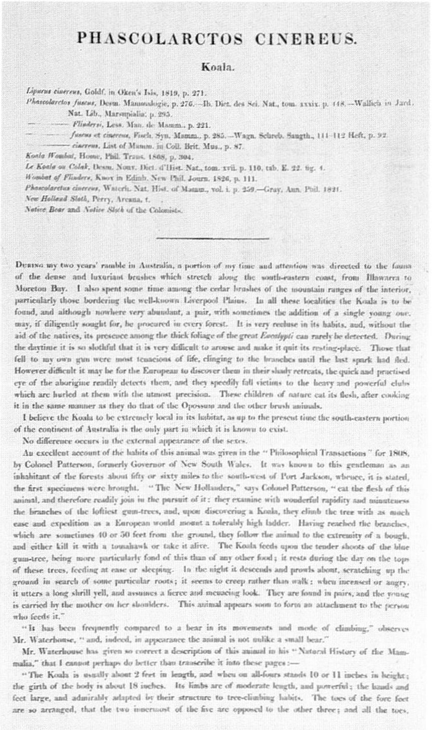

PHASCOLARCTOS CINEREUS.

Koala.

Lipurus cinereus, Goldf. in Oken's Isis, 1819, p. 271.
Phascolarctos fuscus, Desm. Mammalogie, p. 276.—Ib. Dict. des Sci. Nat., tom. xxxix. p. 448.—Wallich in Jard. Nat. Lib., Marsupialia, p. 295.
——— *Flindersi*, Less. Man. de Mamm., p. 221.
——— *fuscus et cinereus*, Fisch. Syn. Mamm., p. 285.—Wagn. Schreb. Saugth., 111-112 Heft, p. 92.
——— *cinereus*, List of Mamm. in Coll. Brit. Mus., p. 87.
Koala Wombat, Home, Phil. Trans. 1808, p. 304.
Le Koala ou Colak, Desm. Nouv. Dict. d'Hist. Nat., tom. xvii. p. 110. tab. E. 22. fig. 4.
Wombat of Flinders, Knox in Edinb. New Phil. Journ. 1826, p. 111.
Phascolarctus cinereus, Waterh. Nat. Hist. of Mamm., vol. i. p. 259.—Gray, Ann. Phil. 1821.
New Holland Sloth, Perry, Arcana, t.
Native Bear and *Native Sloth* of the Colonists.

During my two years' ramble in Australia, a portion of my time and attention was directed to the fauna of the dense and luxuriant brushes which stretch along the south-eastern coast, from Illawarra to Moreton Bay. I also spent some time among the cedar brushes of the mountain ranges of the interior, particularly those bordering the well-known Liverpool Plains. In all these localities the Koala is to be found, and although nowhere very abundant, a pair, with sometimes the addition of a single young one, may, if diligently sought for, be procured in every forest. It is very recluse in its habits, and, without the aid of the natives, its presence among the thick foliage of the great *Eucalypti* can rarely be detected. During the daytime it is so slothful that it is very difficult to arouse and make it quit its resting-place. Those that fell to my own gun were most tenacious of life, clinging to the branches until the last spark had fled. However difficult it may be for the European to discover them in their shady retreats, the quick and practised eye of the aborigine readily detects them, and they speedily fall victims to the heavy and powerful clubs which are hurled at them with the utmost precision. These children of nature eat its flesh, after cooking it in the same manner as they do that of the Opossum and the other brush animals.

I believe the Koala to be extremely local in its habitat, as up to the present time the south-eastern portion of the continent of Australia is the only part in which it is known to exist.

No difference occurs in the external appearance of the sexes.

An excellent account of the habits of this animal was given in the "Philosophical Transactions" for 1808, by Colonel Patterson, formerly Governor of New South Wales. It was known to this gentleman as an inhabitant of the forests about fifty or sixty miles to the south-west of Port Jackson, whence, it is stated, the first specimens were brought. "The New Hollanders," says Colonel Patterson, "eat the flesh of this animal, and therefore readily join in the pursuit of it: they examine with wonderful rapidity and minuteness the branches of the loftiest gum-trees, and, upon discovering a Koala, they climb the tree with as much ease and expedition as a European would mount a tolerably high ladder. Having reached the branches, which are sometimes 40 or 50 feet from the ground, they follow the animal to the extremity of a bough, and either kill it with a tomahawk or take it alive. The Koala feeds upon the tender shoots of the blue gum-tree, being more particularly fond of this than of any other food; it rests during the day on the tops of these trees, feeding at ease or sleeping. In the night it descends and prowls about, scratching up the ground in search of some particular roots; it seems to creep rather than walk: when incensed or angry, it utters a long shrill yell, and assumes a fierce and menacing look. They are found in pairs, and the young is carried by the mother on her shoulders. This animal appears soon to form an attachment to the person who feeds it."

"It has been frequently compared to a bear in its movements and mode of climbing," observes Mr. Waterhouse, "and, indeed, in appearance the animal is not unlike a small bear."

Mr. Waterhouse has given so correct a description of this animal in his "Natural History of the Mammalia," that I cannot perhaps do better than transcribe it into these pages:—

"The Koala is usually about 2 feet in length, and when on all-fours stands 10 or 11 inches in height; the girth of the body is about 18 inches. Its limbs are of moderate length, and powerful; the hands and feet large, and admirably adapted by their structure to tree-climbing habits. The toes of the fore feet are so arranged, that the two innermost of the five are opposed to the other three; and all the toes,

40

have no tail. Their countenance altogether is by no means disagreeable, but harmless-looking and pitiful. They seem formed for climbing trees, but they are rather slow in motion, and but moderately active. Like many other animals of the colony, they are drowsy and stupid by day, but become more animated at night, and when disturbed they make a melancholy cry exciting pity. The feed upon the tops of trees, selecting blossoms and young shoots; and they are also said to eat some particular kinds of bark. When full-grown, they appear about the size of a small Chinese pig. They are certainly formed differently from every other species of the quadrumana, and it is probable they possess different enjoyments. They are very inoffensive and gentle in manners if not irritated.

(From: *Saturday Magazine*, 31 December 1836.)

Just over a quarter of a century later, in 1863, John Gould described koalas in detail and made the first observations of their habitat requirements.

THE TRADE IN KOALA FUR

Soon after their arrival in Australia the European settlers became aware of koala fur. From the 'Sketches of New South Wales' by William Govatt, which appeared in the *Saturday Magazine* of 31 December 1836, came the following account of an Aborigine catching a koala for its fur.

The native espied the animal perched upon the top of a high tree, quite at home. 'Me catch the rascal directly', said the black, and proceeded first to cut a thin pole about ten feet in length. He next tore a long strip of ropy bark, which he fastened to one end of the pole in the form of a loop or noose, after which he commenced climbing the tree in good spirits, and confident of success. The animal, on observing the approach of his enemy, ascended higher and higher till he reached the very extremity of the leafy bough on the top of the tree, while the native, mounting as high as he could safely go, could but scarcely reach him with his pole. For a long time he tried to get the nose over the head of the monkey, and several times when the native imagined he had succeeded, the monkey, at work with his forehand, would repeatedly tear it off and disengage himself. The poor animal, as he looked down upon his perplexing adversary, looked truly piteous and ridiculous, and we began to think that the black would fail in his attempt.

The native, however, growing impatient and angry, ascended a step higher, till the very tree bended with his weight. He tried again, and having succeeded in slipping the noose over the monkey's head, immediately twisted the pole, so as to tighten the cord. 'Me got him rascal' he exclaimed, as he looked downward to see the best way of descending. 'Come along, you rascal, come, come, come,' he cried, tugging away at the monkey, who seemed unwilling to quit his post. Down they came, by degrees, the black cautiously managing his prisoner, every now and then making faces at him, and teasing him, with great apparent delight and satisfaction with himself. We could not but observe the cautious manner in which he appeared at times to treat the monkey but this caution we soon perceived was very necessary, for when they had descended to where the tree divided into two branches, the black endeavoured to make the animal pass him, so that he might have better command over him. In so doing the monkey made a sort of spiteful catch or spring at the native, but which he cleverly avoided by shifting himself to the other branch

41

42

41. The early European settlers soon discovered that there was a market for koala furs. Various methods and instruments were used to catch koalas including wire snares, poisons, shooting and even, as shown here, tomahawks!

42. The koala furs were nailed to trees or pegged out on the ground to dry before sale.

with great dexterity. At length, however, both the man and the monkey arrived nearly to the bottom of the tree, when the latter, being lowermost, jumped upon the ground, got loose, and having crawled to the nearest tree, commenced ascending again. We seized him by the rump, thoughtless of danger, but soon thought it advisable to quit our hold when the native, now enraged, sprung to his tomahawk, and threw it with such force at the unlucky animal as to knock him clean off the tree!

(From: *Saturday Magazine*, 31 December 1836.)

When white settlers recognised the ease with which they could obtain koalas the trade in koala fur developed in the early colonies. The methods employed by koala hunters were many and varied, but poisoning and wire snaring were preferred to shooting as these means damaged the pelt less.

The following account from McQueen's *Social Sketches of Australia* 1888–1975 graphically describes koala hunting techniques. The term 'opossum' was used in early Australia to refer to koalas.

If (killed) by cyanide, a jam tin of water with this in solution, is placed at the foot of a tree or a nearby hollow log, and the morning shows the agony passed through before death gave the animal release. If (killed) by shooting the acetylene search light brought to view the 'possum or bear crouched peering with light lit, frightened eyes from some outstretched branch, or forked limb, a crash! an horrible thud, and there lies one more to be skinned and its white body slung to the dogs or ants. If snared, trappers place slanting saplings against the likely tree, and arrange on each the deadly wire noose through which the 'possum' will thrust his head coming down. In the early morning, before dingoes and crows have disturbed the carcases the trapper does his rounds to collect the strangled 'possums' and bears. All 'joeys' are torn from the pouches, the young ones being thrown to the dogs, and the more developed ones sometimes, and if alive are liberated for future gain.

(From: *Social Sketches of Australia* 1888–1975 by H. McQueen, Penguin Books, Victoria, 1978.)

Australia was seen as the last great frontier for the international fur trade and, not surprisingly, koala fur was soon popular in many overseas fashion and trade centres.

So substantial was the international trade in koala skins that Lydekker observed in 1894 that

the koala must be an abundant animal, since from 10 000 to 30 000 skins are annually imported into London, while in 1889 the enormous total of 300 000 was reached. The value of these skins now ranges . . . from five pence to a shilling each; and they are mainly used in the manufacture of those articles for which a cheap and durable fur is required.

(From: A *Handbook to the Marsupialia and Monotremata* by R. Lydekker, W.H. Allen & Co. Ltd., London, 1894.)

THE FUR SALES

TO-DAY'S OFFERING

110,000 NATIVE BEARS

Although the open season for killing opossums and native bears closed at the end of last month, large numbers of furs still are in the market. Sales of furs will be held in Brisbane to-day, when the offerings will consist of 135,000 opossum furs, against 159,000 submitted two weeks ago.

The number of native bear furs catalogued for to-day is 110,000, against 98,000 offered two weeks ago, and 23,510 a month ago.

At the last fur sales in Brisbane native bear furs brought up to 101/ a dozen for the best specimens, and top grade opossum furs realised up to 164/ a dozen. Average prices, however, were very much below these figures.

Including to-day's catalogues, the total numbers of furs offered at auction as a result of the open season last month were:—

	Bears.	Opossums.
First sale	23,510	51,511
Second sale	98,000	159,000
Third sale	110,000	135,000
Total	231,510	345,511

It is understood that this does not represent the total number of these animals killed for their furs, and that many more are to be marketed.

44

MONEY IN BEARS

£1000 IN ONE MONTH

FIVE TRAPPERS FINED

ROCKHAMPTON, Friday.—In evidence at the Summons Court this morning it was stated that the Kurkowski brothers—Bruno, William, Werner, and Herbert, aged 24, 21, 20, and 18 respectively—during the month of September got 270 dozen bear and opossum skins on Curtis Island. According to the ranger, Mr. A. K. Williams, the Government valuation is £3 10s a dozen, so that the return to the brothers should be about £1000.

The four brothers and Arthur Murphy were prosecuted on two charges each of killing bears, and opossums during the close season, and unlawfully using an acetylene lamp for that purpose. They pleaded guilty and were fined £5 each.

45

43

43. Hunting was common among the early settlers and served as a supplementary income.

44 & 45. Koala hunting became increasingly unpopular as populations in southern Australia became depleted. Only Queensland continued with occasional 'open seasons'.

46. Poetry to appeal to the heart — part of the effort to stop the koala slaughter.

47. A newspaper report showing the build up of strong support to save the koala.

48. Men and women active in the fight against 'open seasons'.

49 & 50. Newspaper cartoons were very effective in drawing people's attention to the koala issue and the injustice and cruelty of the open seasons'.

THE KOALA'S LAMENT.

(By CON. D.)

How have I stirred the white man's wrath,
or angered his god of fame?
I hear the whispering blue gums say,
"Fashion's your enemy's name."
White men made this a Christian land;
Now it's harder to understand
Why they bow to a god of shame!

I do not hamper the white man's work,
or live on his fields of grain;
But I'm doomed to die the dingo's death,
for his greed and gain.
And the tall gums whisper a sad good-bye—
Your heritage lost, now doomed to die,
For Fashion you must be slain.

Blot out my name from your history books. Don't let your children see
Your dressy coats and bloody gold was won by the death of me.
Let me live in their thoughts as their "Teddy Dear,"
Whom they love so well and had no fear;
That is all now I ask of thee.

I was honoured once. My name you bore,
In letters of gold in your bushfolk lore.
I hear them coming—my story's told—
They barter their soul for tarnished gold.

46

SAVING THE NATIVE BEAR

Deputation's Strong Plea

RECONSIDERATION BY CABINET TO-DAY

The fate of the bear will be decided at to-day's meeting of the Cabinet. Yesterday afternoon there was a strong request made to the Acting Premier (Mr. A. J. Jones) foor a reprieve for the koala; in other words, that the Cabinet decision to allow, during August, an open season for the trapping of bears should not be proceeded with, and the bear thus remain protected.

There are good grounds for saying that the request will be acceded to.

47

A HOLIDAY QUESTION FOR SPORTSMEN.
THE DYING BEAR: *"I am a good Australian. What are you?"*

49

WILL IT BE REPRIEVED?
The Cabinet will to-day come to a decision on the representations against an open season for the trapping of native bears.

50

48

By the beginning of the twentieth century, the fur trade had devastated the koala populations in South Australia and Victoria and numbers were declining in New South Wales. Only in Queensland were koalas still present in substantial numbers. Koalas became extinct in South Australia in the late 1930s, largely because of the fur trade.

The focus of the fur trade moved slowly northward as southern populations of koalas diminished. Public concern increased and koalas were protected by legislation in Victoria in 1898. Koalas gained protection in New South Wales under the Native Animals Protection Act of 1903. However, this still allowed for a continuing skin trade and, in 1908, a total of 57 933 koala skins were exported from Sydney. In South Australia koalas were protected under the Animals Protection Act of 1912.

In Queensland open seasons on koalas were declared in 1915, 1917 and 1919. The season from 1 April to 30 September 1919 apparently yielded one million koala skins.

Following public outrage over the 1919 koala open season, the Queensland Government passed the Animals and Birds Act in 1921. This Act was subsequently amended in 1924 to ensure all royalties accrued through its implementation were used to administer the Act.

In December of 1921, the Commonwealth Government introduced controls relating to the export of koala skins through the 'Proclamation Relating to the Exportation of Certain Mammals and the Skins Thereof'.

The reprieve for koalas, however, was short-lived. In 1927, the Queensland Government responded to massive unemployment and economic depression by declaring an open season of one month. Licence fees were received from nearly 10 000 trappers and, during the month of August, 584 738 koalas and 1 014 632 possums were taken. Marshall (1966) records that the total value of skins was £378 023 with koala skins averaging 56 shillings 9 pence per dozen and possum skins 59 shillings per dozen. There were thirty-eight companies involved in

51. During the final 'open season' in Queensland in 1927, nearly 600 000 furs were traded. This, now famous, photograph shows a truckload of 3 600 koala skins collected by a group in the Clermont District of Queensland.

52. Koala furs in a warehouse in Brisbane in 1927. At one time there were thirty-eight companies involved in fur trading, with most furs going to the United States of America.

53. An albino koala.

51

52

fur trading and most of the skins were exported to St Louis in the United States of America.

Confronted by an outraged community and declining koala numbers no further open seasons were declared in Queensland.

It is not possible to tally the total number of koalas killed for the fur trade. Detailed records were not maintained and, in order to sell their koala pelts during closed seasons, trappers often traded koala pelts as 'wombat' fur. The export of koala fur under the guise of other species was apparently brought to an end when President Hoover, a former worker in the goldfields of Western Australia, prohibited the import of koala and wombat skins into the United States of America.

53

ALBINO KOALAS

Albinism, the condition where pigment may be lacking from the skin, hair or eyes, occurs in humans about once in every 20 000 births. It is not known how often albino koalas are born but every few years one is reported. The fur of an albino koala was undoubtedly a prized possession for the early fur traders.

HIS FIRST SNAKE

THE VIRGIN FOREST

A WELCO SHELT

A HOME IN THE FOREST

CUTTING UP

STACKING

GRUBBING

THE CHANGING DISTRIBUTION AND ABUNDANCE OF KOALAS

At the time of European settlement in Australia, koalas were probably common throughout the broad band of eucalypt forests extending from north Queensland to the south-east corner of South Australia. Although possibly restricted to the 'luxuriant brushes', as John Gould reported in 1863, there seems little doubt that koalas were present in large numbers.

As the settlers began to push inland, koalas may have become more abundant. Several early reports suggest that koala numbers were depleted in areas supporting large Aboriginal communities. In 1948, Parris wrote that on the Goulburn River in central Victoria, 'there were no bears . . . when the white men arrived . . . because they were an easy meal for an Aborigine'. By 1870 there were apparently 'thousands [of koalas] on the red-gum timber of the Goulburn', and Parris suggested that this was because the Aborigines had moved.

As the number of settlers grew, it was inevitable koalas would become a target for hunters and the centre of a growing fur trade. The new settlers brought with them axes and saws and, as they pushed further inland, usually along the watercourses, the sound of falling trees began to echo through the valleys. The land they were clearing was to become the lifeblood of the new nation but it was also the forest home of koalas.

As previously mentioned, during the late nineteenth and early twentieth centuries the fur trade was responsible for the death of several million koalas. Between 1887 and 1889, and 1900 and 1903, diseases also significantly reduced their number and distribution. Pathology reports from this time strongly suggest that the diseases then ravaging koala populations and causing their death in the millions were similar, if not the same, as those suffered by koalas today. Reports of local extinction due to disease were common. According to Mr T. Bray, honorary ranger from Vycham Station, Eugowra, koalas were once common along the Lachlan River, in the Forbes District of central New South Wales, but disappeared around 1902 as a result of a disease.

54. The early pioneers moved inland from the coastal settlements and began to clear land to provide timber for building and areas for farming.

55 & 56. As inland towns and villages developed, land clearing increased in scale, in many cases encouraged by government subsidy. The growing road network assisted the vital timber industry.

56

55

57 & 58. Koala populations in southern Australia were devastated by hunting, disease and land clearing during the late 1800s and early 1900s. From 1923 wildlife authorities began relocating koalas from the overstocked islands of Western Port Bay to mainland sites in Victoria and Kangaroo Island, South Australia.

57

By the late 1930s, koalas were considered extinct in South Australia. There were apparently only hundreds in New South Wales, thousands in Victoria, and but ten thousand left in Queensland. While the accuracy of these estimates is uncertain, they give an indication of the extent to which koalas were decimated by the fur trade, disease and the clearing of forests for grazing and cultivation.

When the fur trade was finally brought to a close by protective legislation introduced by State governments and the imposition of export controls by the Commonwealth Government, koala populations began to slowly increase in many parts of their range.

The recovery of koalas in Victoria can be attributed largely to the fortuitous actions of local residents. In the late 1800s, local people transported koalas from the mainland onto Phillip Island and French Island in Western Port Bay. Thriving koala colonies soon developed on both islands. Koalas became so numerous that in 1923 it was necessary for wildlife authorities to relocate some of them onto the mainland.

So successful were these island colonies that several thousand koalas have been subsequently translocated to sites all over Victoria where they were once thought or known to have lived. Thanks to a few caring individuals almost a hundred years ago, koalas are now re-established in many parts of Victoria.

58

Koalas were also successfully reintroduced into South Australia in 1923 when six koalas from Victoria were released into an enclosure on Kangaroo Island. This colony thrived and formed the basis for several koala translocations to mainland areas of South Australia.

Little information is available to document the apparent recovery of koalas in New South Wales from the ravages of hunting and disease around the beginning of the twentieth century. A koala survey carried out in New South Wales in 1949 revealed that koala numbers had increased considerably from the estimate of 200 provided by 'informed naturalists in 1939. In 1963, Mr Allen Strom, Chief Guardian of Fauna in New South Wales, suggested there were several thousand koalas in that State.

Despite the heavy toll of koalas in Queensland during the last open season in 1927, they were apparently still common in many parts of their range immediately after the season closed. Anecdotal reports suggest that koalas in Queensland suffered a dramatic decline due to an epidemic after 1927. The extent of the decline is unknown as it was not until 1960 that any further consideration was given to the distribution of koalas in Queensland. Information collected in Queensland since 1960 indicates that koalas still exist in most areas of suitable habitat.

The distribution and abundance of koalas in Australia have varied greatly since the arrival of the first Europeans. In general, the number of koalas has decreased, initially from hunting and outbreaks of disease and subsequently through the cutting down of the once expansive eucalypt forests of eastern Australia. It is a tribute to their resilience and adaptability that koalas continue to occupy large parts of their original range.

KOALA FACTS

Koalas are only found living in the wild in the four eastern States of Australia — Queensland, New South Wales, Victoria and South Australia — and the Australian Capital Territory.

- They are basically solitary animals, living within a favoured area or home range normally less than 3 hectares in size.
- Male koalas live for about ten years, with females surviving for up to five years longer.
- Adult koalas can weigh between 4 and 14 kilograms depending on their sex and where they are from. Males are up to 50 per cent heavier than females. Koalas from Victoria are normally 8 to 12 kilograms and those from Queensland between 5 and 7 kilograms.
- Koalas spend as much as nineteen hours of every day sleeping.
- Their lethargic lifestyle enables them to survive on the relatively low-energy diet of eucalypt leaves.
- Although they prefer the leaves of just a few eucalypts, koalas have been seen eating or sitting on more than 120 different kinds of eucalypts and nearly forty non-eucalypt tree species.
- In southern Australia the major food tree for koalas is Manna Gum (*Eucalyptus viminalis*) whereas in Queensland it is the Forest Red Gum (*Eucalyptus tereticornis*).
- Koalas get their water from rain droplets, moisture on leaves and by eating leaves. They occasionally drink from streams and ponds and have even been seen swimming.
- Koalas will move several kilometres in search of mates or more tasty food.
- Koalas communicate with one another by marking trees with scent and through calls such as bellows, snarls and screams.
- The breeding season coincides with the warmer spring and summer months.
- Birth occurs thirty-five days after successful mating.
- Although twins are occasionally reported, a single young is the most common.
- At birth young koalas weigh about 500 milligrams and measure less than 2 centimetres from head to tail.
- Cubs are weaned when twelve months old but remain with their mothers for a further twelve months before moving away and establishing their own home range.

59

59. Koalas spend as much as nineteen hours per day sleeping or resting.

60

60. Koalas are at times surprisingly alert and watchful.

61. Unlike most other tree-dwelling animals koalas are generally slow and deliberate in moving around trees.

62. Koalas adopt different positions in the trees depending on how warm or cold it is. Their paws are equipped with rough pads to help them grasp smooth bark and the unusual arrangement of digits on the front paws provides them with a powerful pincer-like grip.

63. Little is known of how well koalas can hear and see although it is believed they have a highly developed sense of smell.

LIFE IN THE TREETOPS

Koalas are uniquely adapted to life in the trees. Unlike possums, gliders and other tree-dwelling marsupials, koalas do not shelter in hollows or nests but rely instead upon their thick, insulating fur for protection from temperature extremes. When faced with cold weather koalas keep warm by curling into a ball and during hot weather they keep cool by dangling their limbs over branches.

Possums and gliders move swiftly and freely about the forest canopy, assisted in climbing and balancing by their long, grasping tails. In contrast, koalas are slow-moving and have small tails. They spend many hours sitting in tree forks and rely upon their strong limbs rather than their tails to move around the treetops. Comparatively long limbs, specialised paws and sharp, pointed claws enable koalas to cling to the smooth branches of eucalypts.

Koala digits are arranged differently on the front and hind paws although all have rough pads to aid their grip. On the front paws the first two digits are opposed to the other three forming a powerful, pincer-like clamp. On the hind paws, only the first digit, which curiously has no claw, is opposed to the others. The second and third digits are fused for most of their length and the dual claws are used for grooming the fur.

Nocturnal animals typically have well developed vision, hearing and sense of smell to help locate food and avoid predators. While little is known of the visual and auditory capacities of koalas, the sense of smell is highly developed and used in selecting suitable food and by males in locating reproductively active females.

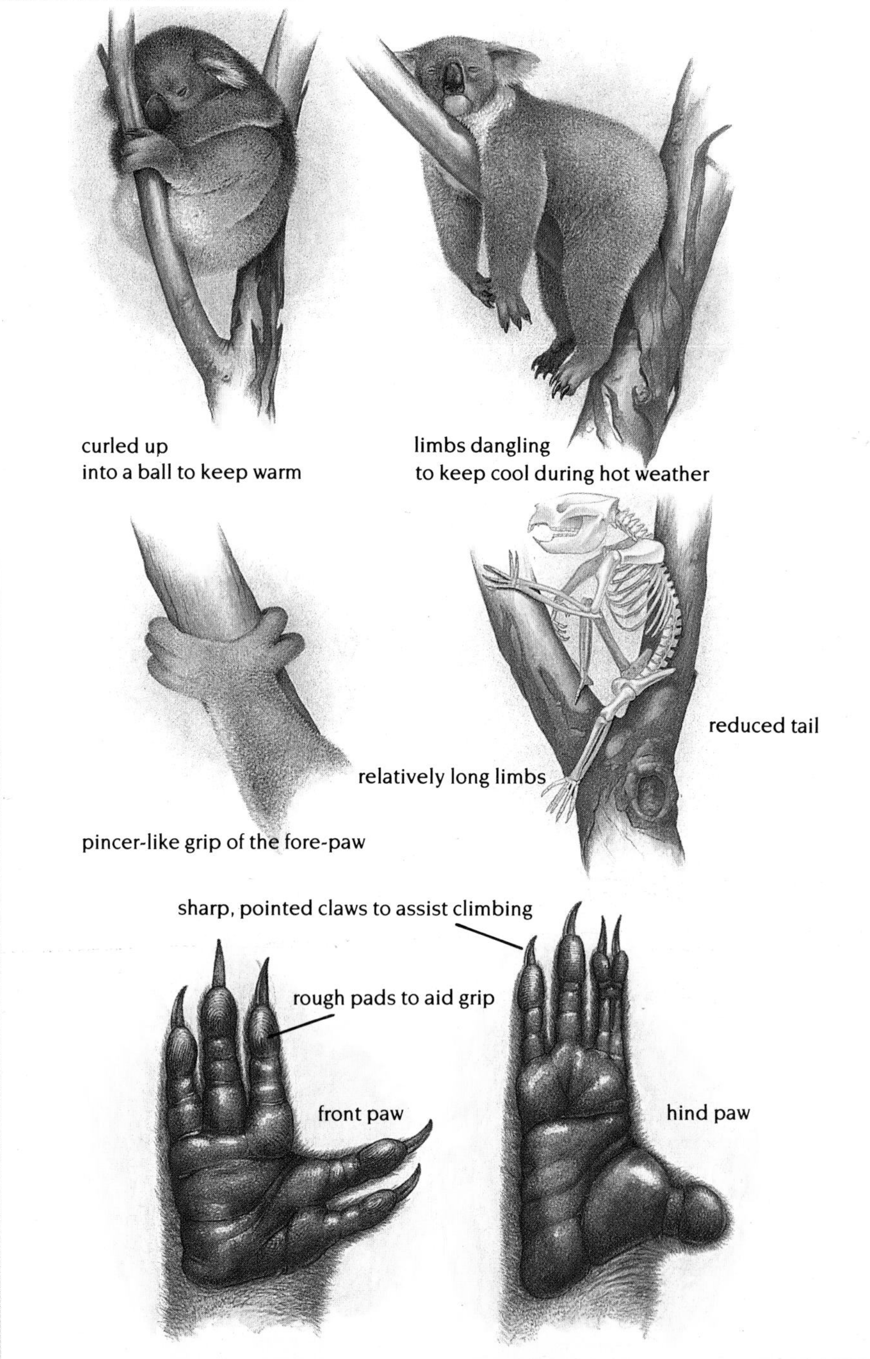

61

62

63

KOALAS TODAY

Although koalas have survived the two centuries since western civilisation came to this continent it is apparent that in modern Australia they face an ever increasing number of threats to their existence. Recent claims that koalas will face extinction by the end of this century have been widely refuted by experts, but the general public is nevertheless concerned. In response to their concern the Koala Conservation Program was established in 1985 and as part of this program a national survey of koala populations was undertaken.

THE NATIONAL KOALA SURVEY

The need for a comprehensive survey to determine the current distribution, conservation status and habitat needs of koalas, was given highest priority under the Koala Conservation Program. This was the unanimous decision in late 1984 of the group of koala experts, veterinarians and State wildlife officers assembled by the Australian National Parks and Wildlife Service to discuss koalas. Until this time no national survey had been attempted, although some State and regional surveys had been carried out before.

With the assistance of $200 000 donated to the Australian National Parks and Wildlife Service in 1985 by American Express, a national survey of koalas was set up to pinpoint locations where koalas warranted special attention. The survey also provided an opportunity to make an informed assessment of the conservation status of the species on a national basis and to determine whether koalas really were on the path to extinction. Apart from describing the distribution of koalas, the survey also aimed to collect details of the type of habitat and tree species occupied by koalas, the dominant land use of the surrounding country and the prevalence of koalas showing disease symptoms resulting from infection with the bacterium *Chlamydia psittaci*.

The Australian National Parks and Wildlife Service coordinated the survey at the national level and the Commonwealth Community Employment Program provided funds to allow State coordinators to be appointed in South Australia, Victoria, New South Wales and Queensland. These people worked with the State wildlife authorities. They distributed survey forms, which were provided by the Australian National Parks and Wildlife Service, to interested persons and groups as well as personnel from government agencies involved in nature conservation and land management. State coordinators were responsible for providing expert advice to those involved in the survey. A training workshop for the State coordinators was held in Melbourne in April 1986. Koala experts, Mr Roger Martin and Dr Kath Handasyde of Monash University, demonstrated practical koala survey techniques during a field excursion to the Brisbane Ranges near Melbourne.

In each State, the coordinators organised koala surveys in key areas and collated and verified survey forms returned to them before forwarding them to the Australian National Parks and Wildlife Service for analysis. Apart from the Commonwealth Community Employment Program coordinators, each State received grants from the Australian National Parks and Wildlife Service/American Express Koala Conservation Fund to assist in financing the survey. Funds were provided to the four eastern States with small amounts also going to the Western Australian Department of Conserva-

65

64. Despite their size, koalas are mostly difficult to find in the forest. They are rarely as obvious as this!

65. Koala survey forms were distributed across the country.

tion and Land Management and the Australian Capital Territory Department of Conservation and Agriculture for reports on their respective koala populations.

The National Koala Survey was undertaken during the 1985–86 and 1986–87 financial years although most surveys took place during 1986 and early 1987.

Distribution of Survey Forms

The distribution of the forms for the national survey was orchestrated through a wide range of government and non-government organisations involved or interested in conservation and land management. Media releases publicised the survey in each State and the coordinators organised and encouraged surveys in key areas. In the rural areas of New South Wales and Queensland survey forms and cards were distributed through suitable newsletters and magazines. Community groups played a major role in organising surveys and recruiting further support from the general public.

Interpreting the Survey Results

The idea of a national survey of koalas conjures up images of an army of observers sweeping through the forests recording the details of every koala in the country. In reality koalas are very difficult to survey. Although quite large animals, koalas can be easily overlooked against the backdrop of a leafy forest canopy. In areas of rugged terrain, koalas are especially difficult to observe. Human observers must discern the surprisingly cryptic form of the sleeping koala while also attempting to move through sometimes dense undergrowth and across steep slopes. The mobility of koalas also poses problems for those involved in organising surveys or interpreting the results. Koalas are more mobile than most people appreciate. A koala noted by one observer may turn up a few days later in a spot several kilometres away from the first and be reported by another observer.

No matter how well trained the observers, or how well organised they are in their searching, large-scale wildlife surveys always have limitations. Koala surveying is no exception and before considering the findings of the National Koala Survey it is important to appreciate these limitations so that the results will not be misunderstood.

Firstly, the National Koala Survey did not set out to answer the question, 'How many koalas are there in Australia today?' This would have been impossible to estimate as survey techniques varied greatly from area to area. Furthermore, some habitats were easier to find koalas in than others, and some surveys used over one hundred people while others were the work of one individual. Many reports were incidental sightings whereas others came from meticulously planned searches. Overall, it was impossible to allow for these differences in 'search efficiency' and impossible to estimate how many koalas were reported more than once. Likewise there was no way of estimating the number of koalas which were missed in the counting.

Rather than estimating the total koala population, the primary aim of the national survey was to describe the current distribution of the species at the national, regional and local levels. Such information when compared with that from previous surveys provides an indication of the status of the species, as well as highlighting the areas where conservation measures are more urgently required.

A further aim of the National Koala Survey was to collect information about the habitat needs of koalas. Observers were asked to provide details of the tree species occupied by koalas, the types of forest koalas were found in and the predominant land-use activites of the area surrounding the forest. It should be remembered that such assessments depended on an observer's knowledge of tree species and forest types, and information gathered was possibly biased in favour of areas where koalas were most readily seen, such as along road verges.

The national survey also aimed to collect data on the factors responsible for koala deaths, for instance, whether they were natural or resulting from human activities. This type of information tended to be biased towards the more obvious causes of mortality, such as road kills and attacks by domestic pets, and is therefore unlikely to truly reflect the proportion of koalas which die from natural causes such as old age.

Participants in the National Koala Survey were also asked to report any sightings of koalas exhibiting the symptoms attributed to infection with the bacterium *Chlamydia psittaci*. The best known symptoms are conjunctivitis and the so-called 'dirty

tail' or 'wet bottom'. While these symptoms are readily recognised in some animals, they are not always obvious during the early stages of disease expression. Equally, not all koalas show disease symptoms despite infection with *Chlamydia psittaci*. For these reasons great care must be taken in interpreting the results for this part of the survey.

For the koala surveys in New South Wales and in parts of Queensland, participants were asked to assess the status of their local koala population by indicating whether they thought the number of koalas in the area had declined, remained stable or increased over the preceding years. Although it was valuable gathering this information, it is important to bear in mind that observer responses may not always have reflected the true status of the local koala population. For example, changes to the dispersal patterns of koalas in an area due to loss of a habitat corridor may be wrongly interpreted as indicating a drop in the koala population rather than a relocation of the focus of the population.

In the following summaries information is presented on the past and present status of koalas in the different Australian States. For the purposes of consistency, all information presented for the National Koala Survey and referred to as 'current' represents the details of koalas reported after 1 January 1985. Most information was collected during 1986 and 1987. The northern and central areas of Queensland were re-surveyed during early 1988.

66

68

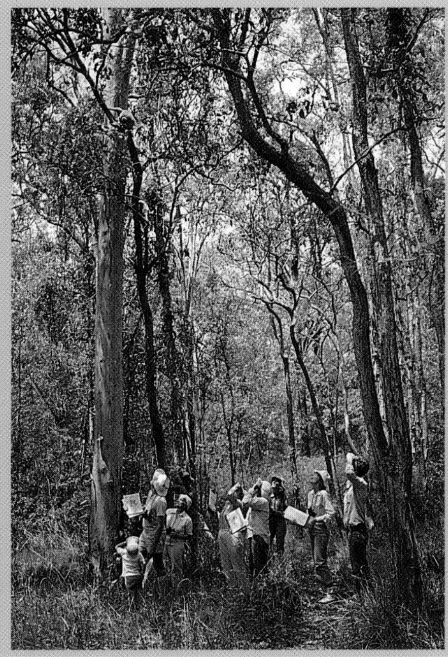

69

66, 67, 68 & 69. Volunteers at work during the koala survey in the Redlands Shire, just south of Brisbane. It was a carefully planned exercise.

THE NATIONAL KOALA SURVEY IN THE REDLANDS SHIRE

In response to the call for assistance with the National Koala Survey, community groups all over the eastern States demonstrated their interest in koala conservation by mobilising their members to undertake koala surveys.

Nowhere in the country was the response more enthusiastic than in the Redlands Shire, about 30 kilometres south of Brisbane on the scenic shores of Moreton Bay. Through State coordinator Ross Patterson, the Koala Preservation Society of Queensland, under the leadership of Mr Don Burnett, generated interest in the survey through extensive media coverage and a number of planning meetings involving church groups, pony clubs, local service groups and representatives of local government. These representatives in turn solicited assistance from personnel within their own organisations and the result was a groundswell of interest in the koala survey.

The weekend for the koala survey was set down for 18 and 19 October 1986, and during the preceding weeks a number of training excursions were organised to assist those involved with recognising koala presence by their characteristic droppings and tree scratchings. 'Tree species identification days' were also organised to help with this aspect of the survey.

The Redlands Shire Council supplied maps of the area, which was divided into survey zones which in turn were allocated to the various groups for them to search.

As the weekend of the koala survey approached, the media campaign advertising the event was intensified and the Koala Preservation Society of Queensland was rewarded by an overwhelming response. Nearly 1 000 people from the local and surrounding areas turned up to help comb the Redlands Shire in search of koalas. As a result, the koala survey carried out in the Redlands Shire was one of the most thorough undertaken in the country and was a tribute to all involved.

67

RESULTS OF THE NATIONAL KOALA SURVEY

KOALAS IN SOUTH AUSTRALIA

History

The early history of koalas in South Australia is poorly documented. Professor F. Wood Jones suggested in 1924 that they were not 'uncommon in certain districts of the South-east of the State' in the period immediately following first settlement.

Koalas first became protected fauna in South Australia in 1912. By this time, the impact of the fur trade and land clearing had already been felt and the number of koalas was severely depleted. Despite their protected status, koalas were unable to recover from the attentions of hunters, and by the late 1930s, were assumed extinct in South Australia.

Introduction of Koalas to Kangaroo Island

The presence of koalas in South Australia today results from the declaration of the Flinders Chase National Park on Kangaroo Island in 1918. Officially reserved as 'a sanatorium and playground for tired workers suffering from brain fag or other forms of overwork' (Dixon 1920), the declaration of the Flinders Chase National Park was the culmination of a program of lobbying, begun in 1888, by naturalists and scientists concerned for the conservation of native flora and fauna.

70

70. The release of the first koalas in Flinder's Chase National Park in 1923. Distinguished anatomist Professor F. Wood-Jones has his back to the camera.

71. Koalas have flourished on Kangaroo Island since their introduction in 1923.

From 1920, management of the Flinders Chase National Park was the responsibility of the Fauna and Flora Protection Board of South Australia. In December 1923, six koalas from Victoria were introduced onto Kangaroo Island. This initial group was released into an enclosure at Rocky River and, in April 1925, there was a further release of six males and six females with young.

The Rocky River koala population quickly became established and escapes from the inadequate wire-netting enclosure were soon common. Koalas moved into the surrounding areas and, apparently, flourished. By the mid 1960s, the koala populations in several areas were so large that serious defoliation of food trees through overbrowsing resulted. The eastern end of Kangaroo Island, near Cygnet River, was colonised by koalas following several relocations from the Rocky River population between 1955 and 1966. Forty-seven koalas were moved to Scotch Thistle Flat, a few kilometres south of Rocky River in 1964. Shortly afterwards koalas were translocated to a number of mainland sites in South Australia.

Reintroduction of Koalas to Mainland South Australia

Koalas from Flinders Chase National Park were translocated to three sites on the Murray River between 1963 and 1965. National Trust Reserves on Goat Island near Renmark, and at Little Toolunka Flat near Waikerie, received four and five koalas respectively during 1963. Two years later a further four koalas from Kangaroo Island were relocated to a Council Reserve at Martin's Bend near Berri.

The colony on Goat Island was established in 1959 with the release of ten koalas from the Adelaide Koala Park. These were the progeny of koalas originally from Queensland and Kangaroo Island. Following the introduction of the additional four koalas from Flinders Chase in 1963, the Goat Island population began to defoliate the trees within their enclosure. It was necessary to increase the size of the reserve and reduce the koala population to five. Three koalas from Goat Island were taken and released at the reserve at Little Toolunka Flat.

All three riverland colonies continued to multiply. Koalas regularly escaped from the reserves, especially after the floods of 1973–74 destroyed the enclosures. By 1975, 'there appeared to be an almost continuous koala population of at least forty animals . . . between Goat Island and Martin's Bend' and around Little Toolunka Flat there was 'a sparsely distributed koala population of at least twenty-four koalas' (Robinson 1978).

Koalas from the Flinders Chase National Park were also released at sites in the Mount Lofty Ranges of South Australia. During 1965, six koalas from Kangaroo Island were released into native bushland on a private property near Ashbourne. Despite occasional sightings, the success of this group in establishing a breeding colony is not known.

71

More successful in establishing colonies in the Adelaide Hills were koalas originally relocated from Flinders Chase into enclosures at Cleland Conservation Park and Belair Recreation Park. According to Robinson (1978) the Adelaide Hills koala population was formed by escapees from these sites in addition to those from an illegal colony of, possibly, New South Wales koalas once located near Brownhill Creek.

In 1969, a private landowner near Lucindale in the south-west of South Australia, and another landowner south-east of Port Lincoln on the Eyre Peninsula, received groups of six koalas for release onto their land. Both groups became breeding colonies and koalas dispersed away from the release sites. By 1975, the colony near Lucindale was estimated to number between twenty and twenty-five koalas. The size of the Eyre Peninsula colony at this time is not known.

Current Status

Since 1975, when the last survey of koalas was undertaken in South Australia, their numbers seem to have decreased in several areas. However, the koala population on Kangaroo Island has continued to multiply and serves as a secure reservoir of animals.

Approximately 1300 person-hours were spent searching the five regions of South Australia in which koalas were known to live. A total of 203 sightings were reported. This is by no means indicative of the total koala population in South Australia today. Several areas known to support koalas were not intensively surveyed due to the difficult terrain. In some areas where the population was believed to range from fifty to 300 koalas, less than 40 per cent were observed and further survey would probably have given increased counts.

Since their introductions onto Kangaroo Island in 1923 and 1925, koalas have thrived and now range over much of the Island. Breeding colonies can be found along most of the river systems where the Manna gum, *Eucalyptus viminalis*, occurs.

As early as 1964, overbrowsing by the expanding koala population was seen as the cause of extensive tree defoliation in the Rocky River area of Flinders Chase National Park. With the growth of the koala population and their dispersal across most of Kangaroo Island, defoliation occurred in a number of areas and is now a serious management problem for the South Australian conservation authorities. Relocation of koalas from Kangaroo Island to mainland sites in South Australia remains an option but few suitable forest areas are available. Translocation of Flinders Chase koalas to mainland areas, in other States, which have more expansive suitable habitats, may provide a solution to the inevitable overcrowding problem on Kangaroo Island.

The decline in koala numbers, since 1975, at a number of locations in the Mount Lofty Ranges, was accelerated by the devastating Ash Wednesday bushfires of 1983. Several koala sites, notably those near Ashbourne and at Horsnell Gully and Cleland Conservation Parks, experienced severe bushfires resulting in many koala deaths. Those that survived the fires were forced to move to nearby localities where the food trees had not been destroyed. At Horsnell Gully Conservation Park, koalas have only recently begun to recolonise the burnt areas that they previously occupied.

The three koala populations, which in 1978 Robinson reported to have dispersed from their initial release sites and established breeding colonies along the Murray River, have apparently declined in numbers. The colonies at Little Toolunka Flat and Goat Island were noticeably reduced and the Martin Bend colony has probably died out. The reason for this is not known.

In the south-east region of South Australia, no koalas were seen near Lucindale in the area where six were released in 1969. The South Australian survey coordinator, Rodney

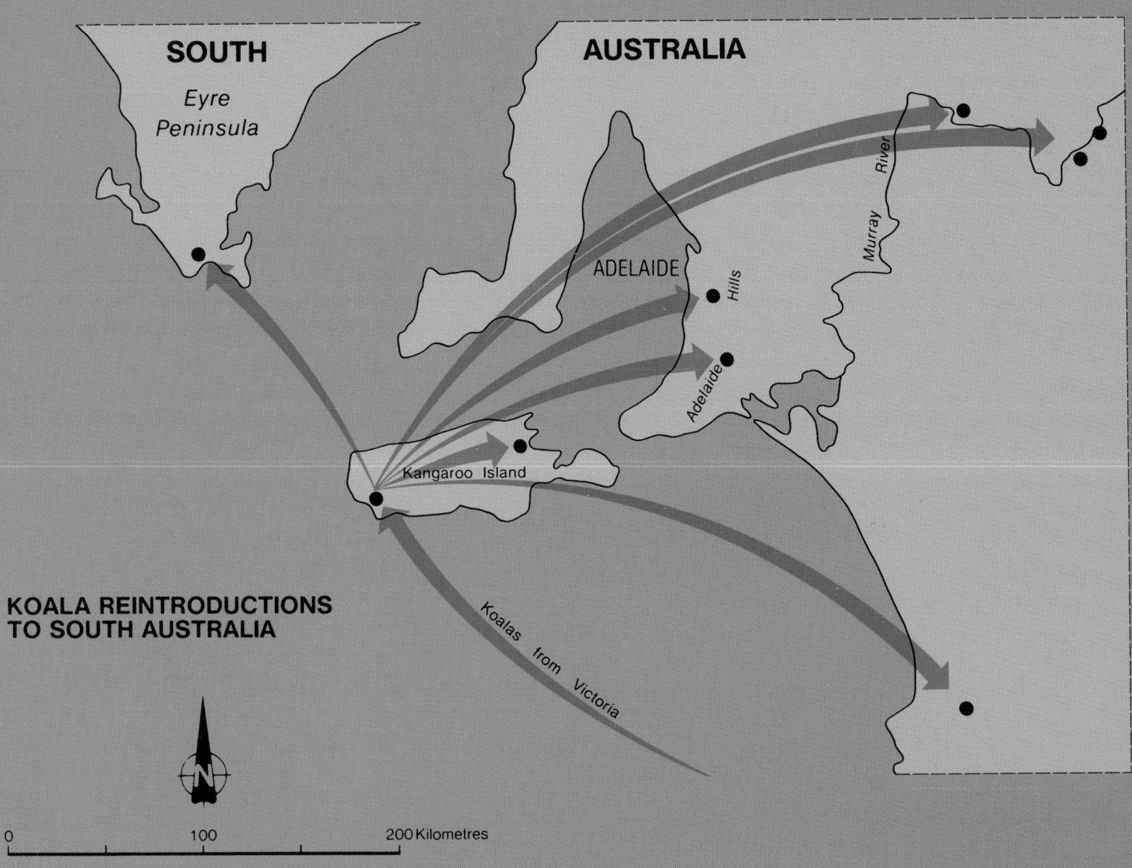

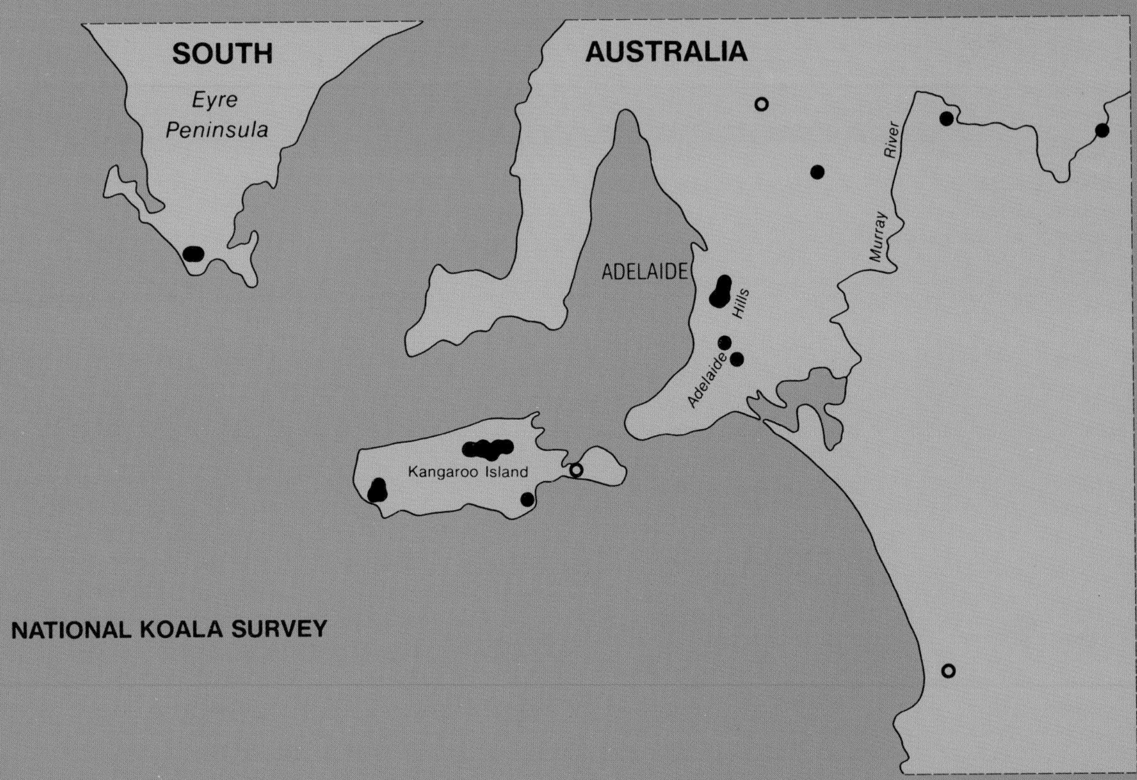

Koalas in South Australia

By the late 1930s koalas were considered extinct in South Australia due primarily to the combined effects of hunters and land clearing. Koalas brought from Victoria in 1923 and 1925 flourished on Kangaroo Island and their descendants were used to recolonise a number of sites on mainland South Australia in the 1960s (upper map). During the National Koala Survey (lower map) koalas were positively sighted (●) at all known sites except near Lucindale in the south-east of the state where only fresh droppings indicated koalas were still in the area (o).

Spark, did find recent koala droppings so some koalas are still present.

The most successful of the mainland translocations of koalas from Kangaroo Island has been to the private bushlands on the Eyre Peninsula. From an initial release of two males and four females in 1969, the colony had grown to approximately forty or fifty koalas in 1986. Tree defoliation has not become a problem as yet and additional areas are available for the colony to disperse throughout as it increases.

KOALAS IN VICTORIA

History

Before the first European settlement of Victoria, koalas were probably widespread and common across large areas of the State. The best account of the history of koalas in Victoria is provided by Martin in his 1983 description of koalas in the Gippsland area of south-east Victoria.

Reports from early explorers in the Gippsland region indicated that koalas were once common. Journeying through the tall, wet forests in April 1840, one of Paul Strzelecki's party reported that they

> saw only one animal through the country we passed, the size of a small dog which lives in trees — a monkey or native bear. We got some by shooting, some by the natives climbing the trees.

(From: *The Count: A Life of Sir Paul Edmond Strzelecki, KMG, Explorer and Scientist* by G. Rawson, William Heineman Ltd, Melbourne 1953.)

Four years later, George Robinson, the Chief Protector of Aborigines, journeyed through the low and more open forests to the south of the route taken by Strzelecki. A member of Robinson's party, Haydon, recorded that

> on passing through some belts of stringy bark forest the black police succeeded in killing five bears called in their language Carbora. They are a species of sloth and are stupid looking creatures never attempting to get away from their pursuers who climb the tallest trees after them.

(From: *Five Years Experience in Australia* by G.H. Haydon, Felix, Hamilton Adams and Co., London 1846.)

There are numerous accounts of Aborigines hunting koalas to provide food for the early explorers of the Gippsland and there is little doubt that koalas formed a regular part of the diet of the Aborigines. Before the Gippsland region was opened to the first European settlers, with the establishment of Mcdonald's Track (from Dandenong to Morwell) in the early 1860s, several explorers remarked on the impact of Aboriginal hunting on koala populations. During his 1844 expedition through the area, George Robinson recorded that

> the forest animals have vastly increased since the destruction of the local inhabitants (Aborigines) . . . the Boonwerong nation . . . by the Gippsland aborigines. The . . . Carbora of the natives, Bear or Monkey of the settlers . . . were in places abundant.

(From: *George Augustus Robinson's Journey into south-eastern Australia, 1844* by G. Mackaness, Australian Historical Monograph vol.19, Review Publications, Dubbo, New South Wales, 1978.)

Martin (1983) suggested that the impact of Aborigines on the Gippsland koala populations may have been greater in the coastal areas than in the taller inland forests. The Kurnai Aborigines, who lived in south Gippsland and were estimated to number nearly 2 000, moved with the seasons, spending spring, summer and autumn in the coastal areas and only the winter further inland. Coutts (1970) recorded that the Kurnai 'probably lived on a diet of fish, birds and shellfish supplemented with occasional marsupials (wallaby and bear) and various vegetable foods'.

Until the white settlers arrived and set about their eradication, dingoes were also a major predator of koalas. Coverdale (1920) stated that dingoes were 'the deadly enemies of the bears, killing numbers of them when the latter came down to change trees'.

The settlement of Gippsland in the 1860s had a two-fold effect. The Aborigines and dingoes were driven out and, consequently, the predation pressure on koalas was reduced. However, any positive effects for the koala populations from the arrival of settlers were short-lived as vast tracts of forest were cleared. To encourage the growth of grasses ideal for grazing livestock, the settlers regularly burnt their cleared land and sometimes, by accident, the adjoining forests were also burnt. A massive forest fire in 1898, which burnt for nearly a month, significantly reduced the Gippsland koala populations which had already suffered from the previous forty years of white settlement.

Wilsons Promontory escaped the major fires of the late 1800s and

KOALAS IN WESTERN AUSTRALIA

History

In Western Australia today there is relatively little forest suitable for koalas. This was not always the case as illustrated by the sub-fossil remains of koalas described by Ludwig Glauert (1910) from *The Mammoth Cave* in the far south-west of the State. Except for recent introductions, koalas have not been found in Western Australia in modern times.

In 1938 one male and three female koalas were introduced into an enclosure at Yanchep National Park, 48 kilometres north of Perth. These koalas, apparently from Victoria originally, were transferred from Perth Zoo which was unable to supply sufficient food. By October 1941, Congreve and Betts (1978) reported that these animals had died.

It was not until 1947 that a further two Victorian koalas were brought to Yanchep. Over the next four years a further eight koalas were added to the colony, two from Victoria and six from Queensland. Further koalas were brought from Kangaroo Island, South Australia, Victoria and Queensland during the period from 1959 to 1975. In 1972 two male koalas and one female were exchanged by Yanchep National Park and Lone Pine Koala Sanctuary in Queensland to avoid problems of in-breeding.

Current Status

At last count there were eleven koalas in the colony at Yanchep National Park. The future of the colony is uncertain as some are infected with *Chlamydia psittaci* and there has been no successful breeding for several years.

accordingly the koala population increased with the decline in the number of Aborigines and dingoes in the area. By the early 1900s, the large koala population had come to the attention of hunters and, in 1906, Hardy reported that 'the demand for skins of wallaby and koala from the Promontory has been so great in the past that over 2 000 of each have been removed in one year'.

Wilsons Promontory was declared a National Park in 1905, thus bringing to an end all hunting activities. The koalas recovered to such an extent that defoliation of coastal food trees became a problem. Although steps were taken to reduce the number of koalas, tree deaths occurred on a large scale, some koalas died as a result and others moved to the taller inland forests.

According to Lewis (1954), by 1925, the Victorian koala population was thought to number 'about 500' and many naturalists and conservationists expressed fears that koalas were headed for extinction in southern Australia.

Fred Lewis, Chief Inspector of the Victorian Department of Fisheries and Game for thirty-three years stated in 1934 that

> on the mainland of Victoria, I feel certain, the koala is doomed to early extinction, and will never be re-established, excepting perhaps in some reserves which may be specially set apart for its protection and conservation.
>
> (From: *The Koala in Victoria*, by F. Lewis, Victorian Naturalist, vol 51, 1934.)

The Islands of Western Port Bay

The islands of Western Port Bay have played an important role in reestablishing koalas throughout large parts of their previous range in Victoria and South Australia. Although the islands possessed habitat suitable for koalas, they were not present there until the late 1800s when local farmers and fishermen relocated koalas from the mainland onto Phillip and French Islands. These citizens could not have known the significance of their actions.

The first group of koalas were released onto Phillip Island in 1870 by Mr J. Smith of Bass River. Shortly

72

73

74

72, 73 & 74. Conservation authorities relocating koalas from Western Port Bay to mainland sites. The first translocations were in 1923 and the program has continued until the present.

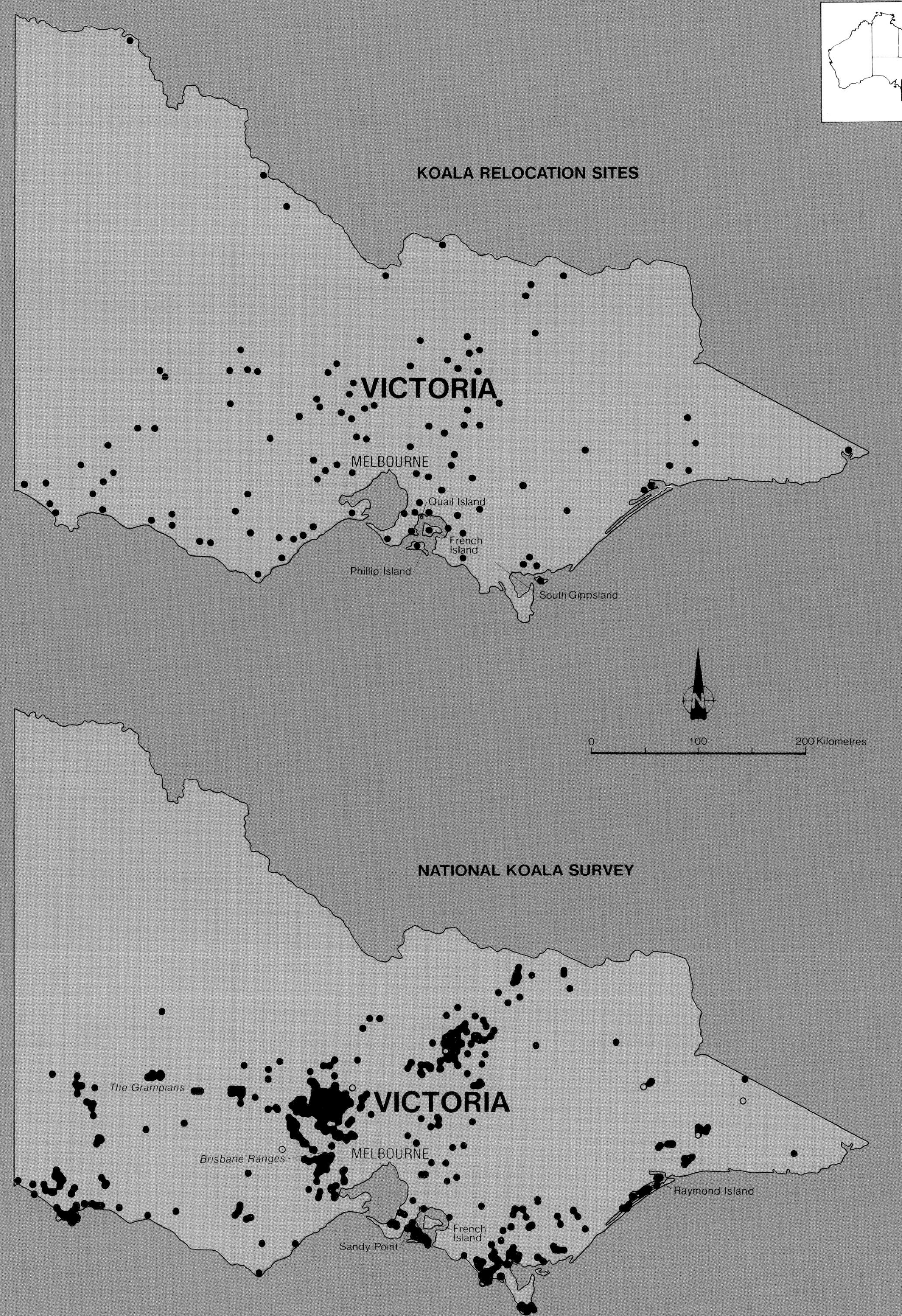

Koalas in Victoria

Koalas were once common throughout large areas of Victoria but by 1925 some authorities believed they numbered around 500. Fortunately koalas released onto French Island and Phillip Island during the late 1800s established successful breeding colonies. The subsequent relocations of nearly 10000 koalas from these overcrowded populations to mainland sites (shown on upper map) has helped to re-establish koalas in many parts of Victoria. During the National Koala Survey (lower map), koalas were observed (●) or their presence indicated by characteristic droppings, calls and scratch marks on trees (o) at most of the sites where koalas have been released over the past 65 years.

after, several others followed Smith's example and released small groups of koalas from Flinders and Gippsland onto Phillip Island. About ten years later, the local fishermen of Corinella also relocated some koalas from the mainland, this time onto French Island. Breeding colonies were soon established on both islands.

The French Island koalas multiplied more rapidly than those on Phillip Island and, by 1920, the island farmers complained of tree defoliation by the koalas. Consequently, in 1923, fifty koalas were translocated from French Island to Phillip Island. The introduction of new animals to Phillip Island acted as a catalyst and this population began to increase steadily. In 1923 and 1925, koalas were also removed from French Island and transferred to Kangaroo Island, South Australia.

Despite the removal of koalas to Phillip Island and Kangaroo Island, by the early 1930s, koalas on French Island had multiplied to such an extent that eucalypt defoliation became a major problem. To ease the burden on the trees of French Island, 165 koalas were translocated to nearby Quail Island between 1930 and 1935.

The Quail Island koala population soon became a thriving breeding colony. On this relatively small island, massive defoliation of the trees was noted by 1943 and large numbers of koalas died due to starvation. In 1944, wildlife authorities removed the total population of 1349 koalas from Quail Island and released them at various sites on mainland Victoria. Quail Island remains free of koalas today.

With the large-scale removal of koalas from Quail Island, a translocation program had become established and has continued almost annually ever since. Between 1923 and 1987, over 5 000 koalas were translocated from French Island. Except for those used to establish or supplement the populations on Phillip, Kangaroo and Quail Islands, these koalas have been released at mainland sites.

After the release of koalas from French Island onto Phillip Island in 1923, this population began to increase more rapidly than before and by the late 1930s, overbrowsing of the food trees was causing concern to the local authorities. In 1941, it was considered necessary to remove koalas from Phillip Island in order to preserve the eucalypt trees. Between 1941 and 1978, when translocations from Phillip Island were halted (due to a decline in the population attributed to disease and increasing road tolls), Backhouse and Crouch (1988) reported that nearly 3 200 koalas were removed and released at mainland sites.

Sandy Point, within the naval base at HMAS *Cerebus* on the western side of Western Port Bay, was one mainland site where koalas from French Island were relocated. In 1972, twenty koalas were released and by 1984 the koala population was estimated at 345. Concerns about overbrowsing were already being expressed at this time and translocations of koalas from Sandy Point to other mainland localities began in 1985. A total of 203 koalas were taken from the Sandy Point colony during 1985, 1986 and 1987.

Current Status

The distribution of koalas in Victoria today largely reflects the program of translocations from the islands of Western Port Bay which began in 1923. Nearly 10 000 koalas have been relocated under this program which has re-established koalas throughout much of their previous range in Victoria. Up to the end of 1988, koalas had been translocated to nearly 150 mainland sites across the State of Victoria. Although little effort has been made to determine the success or otherwise of these releases, the findings of the National Koala Survey indicate that koala populations are still present in most of the areas where liberations have taken place.

In 1978 Warneke reported that the Victorian Department of Conservation, Forests and Lands (and formerly Victorian Fisheries and Game Department) had operated for a number of years under the policy that koalas should only be released into large areas (minimum area of 4 000 hectacres) of suitable forest, with guaranteed tenure and established management practices. The success of koala re-establishment throughout Victoria strongly vindicates this policy.

In total, 521 survey forms were completed during the National Koala Survey in Victoria. Of these, twenty-four reported no evidence of koalas. The remaining 497 forms as well as incidental reports recorded positive sightings of 1 702 koalas at 696 localities. At a further twenty-seven sites the presence of koalas was inferred from their characteristic vocalisations, droppings and scratch marks on trees. Reports of koalas found dead were received from fifty-three localities representing sixty-three koalas in total. Thirty-nine (61.9 per cent) of these mortalities were attributed to collisions with motor vehicles.

The number of koalas observed during the national survey in Victoria should not be considered as an estimate of population size. For the reasons discussed previously, it is not possible to gain such an estimate from the National Koala Survey. The best indication of population status which a survey of this type can provide is given by the number and distribution of sites where the species is observed. It is encouraging that for Victoria koalas were reported from nearly 700 sites ranging across those areas of the State which retain suitable forest habitat.

From being perilously close to extinction in Victoria just seventy years ago, koalas have made a remarkable recovery, albeit with a little help from interested persons and conservation authorities, and can now be considered widespread across the State and even common in the relatively few areas of ideal forest habitat which remain.

KOALAS IN NEW SOUTH WALES

History

The history of koalas in New South Wales is poorly recorded. Like koalas in other parts of Australia, those in

75. Publicity and questionnaire for the 1967 New South Wales koala survey.

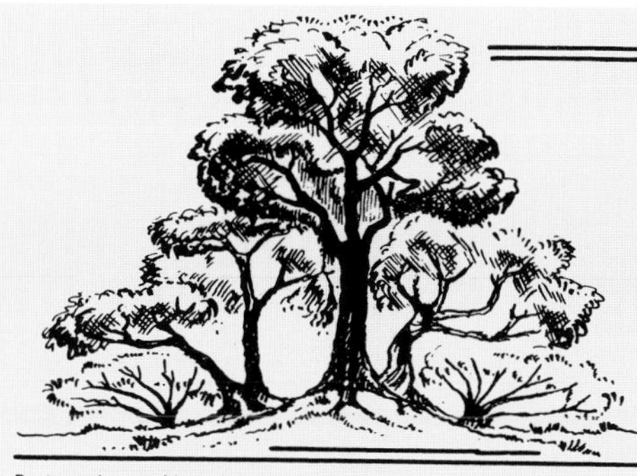

The Junior TREE WARDEN

Registered as a Newspaper — No. 1 — 1967 — PAGE 1

A QUARTER MILLION CHILDREN TO MAKE ENQUIRY

HELP FOR KOALAS

The New South Wales Wildlife Service is to undertake the most extensive survey of koala distribution yet attempted in New South Wales.

The area supporting this delightful marsupial which depends upon a relatively small number of eucalypts and angophoras is at present a mystery. In the past two years some 40 sightings have been reported from widely separated locations in eastern New South Wales.

With the present survey the Wildlife Service hopes to learn more of the koala's whereabouts, its food trees and also gain some idea of numbers.

"Because it is perhaps the most difficult of all the larger marsupials to see, so the more eyes looking, the better," explained Mr. Strom, Chief of the Wildlife Service. "For this reason, we are enlisting the aid of a quarter of a million pairs of eyes through the 'Junior Tree Warden'," he added **(see Page 8).**

During the first week in August all eyes should be turned treewards in an effort to see this elusive creature.

1967 KOALA SURVEY NEW SOUTH WALES

School .... Teacher .... Class ....

OBSERVATIONS DURING CONSERVATION WEEK, 1967

1. We have seen wild koalas this week. Yes/No*
2. We have seen [1] .... koalas this week.

OBSERVATIONS SO FAR IN 1967

1. We have seen wild koalas this year. Yes/No *
2. We have seen [1] .... koalas this year.

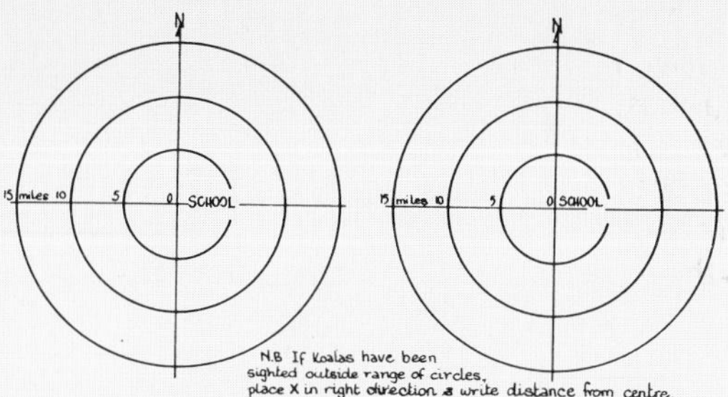

3. Mark map with cross where koalas were seen.
4. We have forwarded a sample of the koala's tree as suggested on P.7 to the Wildlife Service, A.D.C. House, Kent Street, Sydney.

3. Mark map with cross where koalas were seen.

FROM THE COMMENTS OF "OLD-TIMERS"

1. Koalas were found within 15 miles of the town 25 years ago. Yes/No*.
2. Koalas were found within 15 miles of the town 50 years ago. Yes/No.*
3. Koalas were found within 15 miles of the town before 1900. Yes/No.*
4. About what year was the last koala sighted? ....
5. Have there been any large bushfires within 15 miles of the town? Yes/No.*
6. What years were they? ....
7. Have koalas been seen with any disease within 15 miles? Yes/No.*
8. Was there a time when koalas were shot or killed?
9. When? ....

* Strike out which does not apply.
[1] Write number.

NOTE. If this paper does not reach you prior to Conservation Week, then work on it the first week of September or October. All to be posted by September 15th or by October 12th.

75

New South Wales suffered greatly from the arrival of European settlers. Land clearing, hunting and disease took such a toll on New South Wales koalas that by the late 1930s they were reported to number only a few hundred.

In 1949, the Fauna Protection Panel undertook a koala survey in New South Wales. Letters were sent to land inspectors, field officers of the Department of Agriculture, Forestry Officers and Pasture Protection Boards. The *Education Gazette* was used to inform school teachers of the survey and all police stations were also notified.

As a result of the 1949 survey 109 koala colonies were identified. It was concluded that although the koala was once 'more widely spread than it is today' there 'appear to be some thousands of koalas at present in the state' (only known copy of the report held by V. Serventy). The report from the survey further recorded that 'the greatest density [of koalas] occurs on the far North Coast as there were . . . reports from 26 localities north of the Clarence River'. Few koala colonies were identified from south of Sydney, one notably from the Snowy River area.

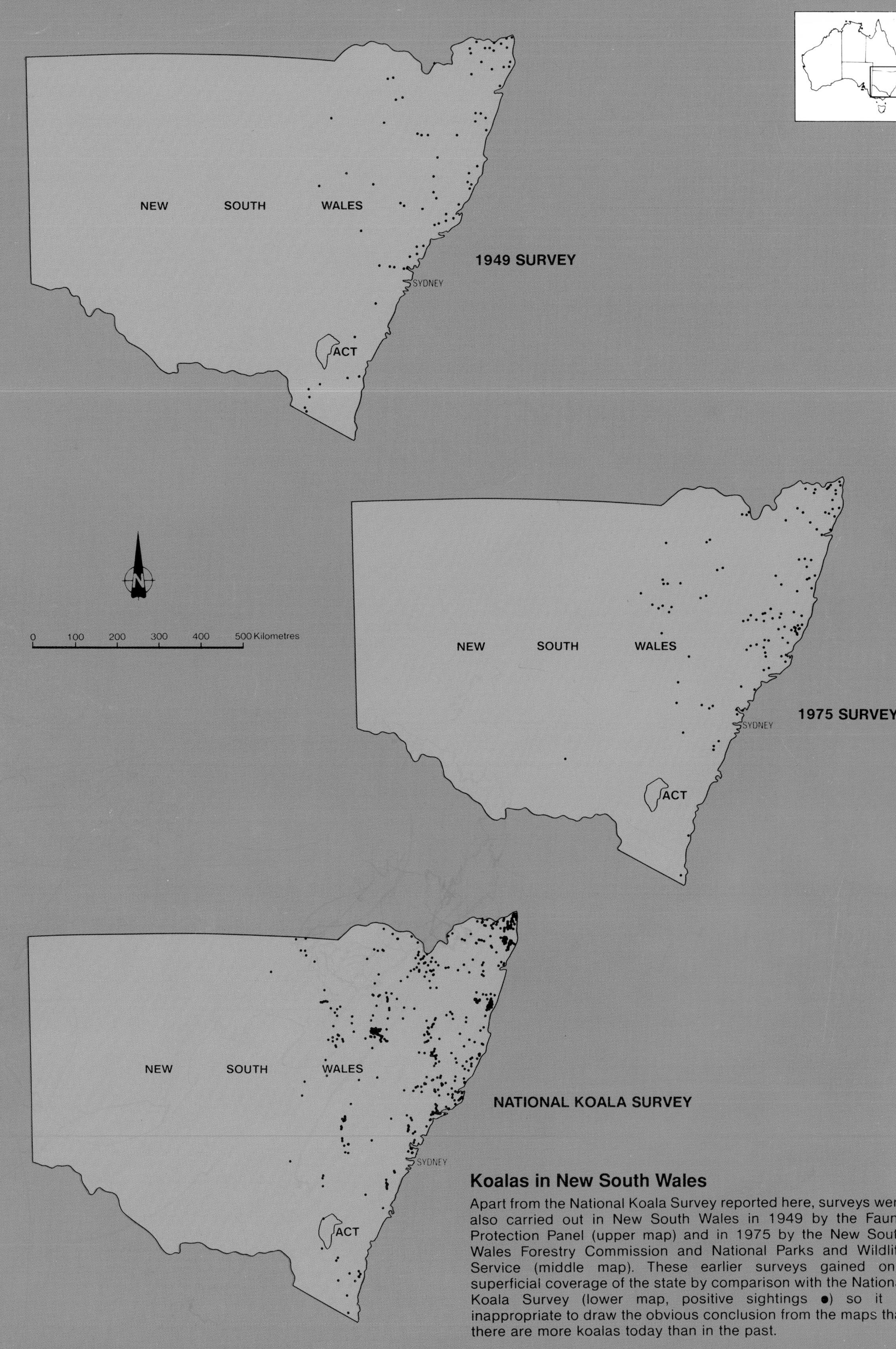

Koalas in New South Wales

Apart from the National Koala Survey reported here, surveys were also carried out in New South Wales in 1949 by the Fauna Protection Panel (upper map) and in 1975 by the New South Wales Forestry Commission and National Parks and Wildlife Service (middle map). These earlier surveys gained only superficial coverage of the state by comparison with the National Koala Survey (lower map, positive sightings ●) so it is inappropriate to draw the obvious conclusion from the maps that there are more koalas today than in the past.

76 & 77. Two koalas, showing their typical disregard for prying human eyes.

76

77

In 1967 a further koala survey was carried out by the then newly formed New South Wales National Parks and Wildlife Service. The survey questionnaire was distributed to all government schools in New South Wales through the newsletter of the Tree Warden's League of New South Wales. Sadly, much of the information which this survey yielded has been lost, so little is known of the status and distribution of koalas at this time.

It wasn't until 1975 that another attempt was made to survey the koala populations of New South Wales. In a cooperative effort the New South Wales Forestry Commission and National Parks and Wildlife Service distributed a questionnaire to all regional offices seeking details of koalas found within the areas under their control. The survey specifically requested details of koalas observed since 1 January 1970. Gall and Rohan-Jones (1978) reported that the survey produced 353 reports of koala sightings up to its conclusion in March 1976.

As was the case for the 1949 survey, during the 1975 survey the greatest concentration of koala sightings was in the northern coast region of New South Wales, with around 80 per cent of reports coming from this area. Only four sightings were recorded south of Sydney and Gall (1978) concluded that some 'shrinkage of range' had occurred. Despite this observation, Gall was heartened by the finding that koalas were observed in 'about a hundred "secure" areas' and concluded that 'the koala, at least by current standards of status, is common and secure in New South Wales'.

Current Status

The previous koala surveys undertaken in New South Wales were superficial by comparison with the National Koala Survey. Consequently, caution is needed in making comparisons between the findings of previous koala surveys and those of the national survey.

In total, 2 564 survey returns were received in New South Wales during the National Koala Survey. Of these 697 reported positive koala sightings after 1 January 1985. As expected from previous surveys, the overwhelming majority of these reports were for koalas within the north-eastern quarter of New South Wales. Koalas were only sparsely distributed on the western plains and rare on the south and central coast of New South Wales. Given the paucity of koala sightings south of Sydney since the earliest surveys were undertaken, Reed, Lunney and Walker (1988) have suggested that 'the extensive forests in and around the Kosciusko National Park' may have always been 'marginal koala habitat'.

The survey card used in the New South Wales part of the national survey asked the observers to indicate how often they saw koalas.

The commonest response was that koalas were seen 'yearly' (292 or 41.9 per cent of replies), followed by 'intermittently' (213, 30.6 per cent), 'monthly' (128, 18.4 per cent) and 'weekly' (64, 9.1 per cent). Perhaps surprisingly, the 'weekly' and 'monthly' sightings were not concentrated in any particular area. They were scattered across the State suggesting that large concentrations of koalas occur at a number of discrete locations in New South Wales.

Reed, Lunney and Walker (1988) have compared the results of the National Koala Survey with the koala distribution information collected up until 1985. Dividing New South Wales into 299 grid cells, representing all past and present sighting records, they found that during the national survey koalas were not reported from 114 (38 per cent) of the grid cells where they had been observed during previous surveys. In 113 grid cells (38 per cent) koalas were still present and in the remaining seventy-two cells (24 per cent) the first reports of koalas were obtained during the National Koala Survey. The latter 'expansions' to the range of koalas in New South Wales probably reflects the more widespread survey achieved by the national survey rather than any real population growth.

Overall, the findings of the National Koala Survey in New South Wales suggest that despite the loss of some populations towards the southern and western margins of their range, the broad historical distribution of koalas is being maintained. It may be reason for concern that koalas were recorded as abundant in only a few scattered locations. The area where most koalas are found, namely the northern coastal region of New South Wales, seems destined for intensive urban expansion and tourist development over the next decade which clearly will pose a direct threat to the koalas of that region.

78. From the column 'The G.O.M. of Bungendore: An Appreciation by Sir J.M. Carruthers' when reminiscing of the 1870s and 1880s, the *Queanbeyan Age*, 8 April 1919.

79. An excerpt from the Bungendore Column, the *Queanbeyan Age*, 13 July 1883.

80. A.B. (Banjo) Paterson referred to 'native bears' in his well-known work *The Gundaroo Bullock*. Gundaroo is about twenty minutes drive north-east of Canberra.

81 & 82. Koalas have been introduced several times into the Australian Capital Territory. The first time was in 1939. Koalas are seen here being released into the Tidbinbilla Nature Reserve near Canberra.

of the town boundaries. I am sorry for the part I took in the extermination of a fine animal now seldom if ever seen in these parts. In the trees on the flat at the mouth of Deep Creek one could at any time count 100 native bears in the fine timber there—within a few hundred yards from our meal camp. I remember how, in a party of six of us from Sydney, which included a well-known banker, a barrister, and a young solicitor, who afterwards was a colleague of mine in Parliament, we had to rebuke the latter because he spent a morning shooting native bears and killed 31. His excuse was that he wanted a rug and that Murphy was skinning them for him. Fact was that about a dozen were skinned and the rest wasted. The trees around the Lake were then gradually dying out, either as the result of borers or from being sodden at the sub-soil. Some

78

were brought before them, they would inflict the highest penalty.

A party of four persons went bear shooting last week, and in six hours managed to secure 107, which were soon deprived of their skins. The fur is excellent at this time of the year.

Messrs. M'Jannett and Hawks have formed a small party and started for the Murrumbidgee

79

The Gundaroo Bullock

Oh, there's some that breeds the Devon that's as solid as a stone,
And there's some that breeds the brindle which they call the 'Goulburn Roan',
But amongst the breeds of cattle there are very, very few,
Like the hairy-whiskered bullock that they breed at Gundaroo.

Far away by Grabben Gullen, where the Murrumbidgee flows,
There's a block of broken countryside where no one ever goes;
For the banks have gripped the squatters, and the free selectors too,
And their stock are always stolen by the men of Gundaroo.

There came a low informer to the Grabben Gullen side,
And he said to Smith the squatter, 'You must saddle up and ride,
For the bullock's in the harness-cask of Morgan Donahoo —
He's the greatest cattle-stealer that abides in Gundaroo.

'Oh, ho!' said Smith, the owner of the Grabben Gullen run,
'I'll go and get the troopers by the sinking of the sun,
And down into his homestead tonight we'll take a ride,
With the warrants to identify the carcase and the hide.

That night rode down the troopers, the squatter at their head,
They rode into the homestead, and pulled Morgan out of bed.
'Now, show to us the carcase of the bullock that you slew —
The great marsupial bullock that you killed in Gundaroo'.

They peered into the harness-cask, and found it wasn't full,
But down among the brine they saw some flesh and bits of wool.
'What's this?' exclaimed the trooper — 'an infant, I declare,'
Said Morgan, 'Tis the carcase of an old man native bear.
I heard that ye were coming, so an old man bear I slew,
Just to give you kindly welcome to my home in Gundaroo.

'The times is something awful, as you can plainly see,
The banks have broke the squatters, and they've broke the likes of me;
We can't afford a bullock — such expense would never do —
So an old man bear for breakfast is a treat in Gundaroo.'

And along by Grabben Gullen where the rushing river flows,
In the block of broken country where there's no one ever goes,
On the Upper Murrumbidgee they're a hospitable crew,
But you mustn't ask for 'bullock' when you go to Gundaroo.

(From: The *Collected Verses of* A.B. *Paterson*, Angus and Robertson, Sydney, 1921)

80

KOALAS IN AND AROUND THE AUSTRALIAN CAPITAL TERRITORY

History

Koalas were once common in and around the Australian Capital Territory. However, by 1901 they were rare in the district as a result of the combined effects of bushfires, habitat destruction and hunting.

Introduction of Koalas into the ACT

There have been several introductions of koalas into the Australian Capital Territory, the earliest dating from 1939. Most koalas came from French Island and Phillip Island in Victoria and were released into the Tidbinbilla Nature Reserve. During 1978–79 koalas were also released at Bushfold Valley and Orroral Valley in the Namadgi National Park, south-west of Canberra.

The six koalas introduced into Tidbinbilla Nature Reserve in 1939 apparently survived and bred successfully within the enclosure. The boundary fences fell into disrepair in about 1944 and these animals escaped. After the 1970 introduction escapes were common until 1984 when a substantial upgrading of the fences took place. It seems likely that those animals which escaped and their progeny spread throughout the Tidbinbilla Valley and possibly beyond. Two koalas were seen in the Brindabella Range, south-west of Canberra, between 1969 and 1972 and Eberhard and Shulz (1973) reported that another koala was regularly observed in the same area about ten years earlier.

Between 1974 and 1980 koalas were routinely observed outside the enclosure in the Tidbinbilla Nature Reserve. There have been unconfirmed reports of koalas as far away as 20 kilometres and tagged animals have been seen 5 kilometres from the enclosure.

Between 1980 and 1984 there were several sightings of koalas near the Orroral Valley Tracking Station and in 1986 two koalas were seen in the Stromlo Pine Forest.

Up until the introductions of koalas from Phillip Island in 1978 those released into the Australian Capital Territory were ostensibly healthy and had a high reproductive success. The Phillip Island koala population has been infected with the bacterium *Chlamydia psittaci* since at least 1954. As a consequence of introducing these koalas into the Australian Capital Territory populations, Braysher, of the local Parks and Conservation Service, considers that the breeding success of koalas in the enclosure and probably in the wild, has been significantly reduced.

81

82

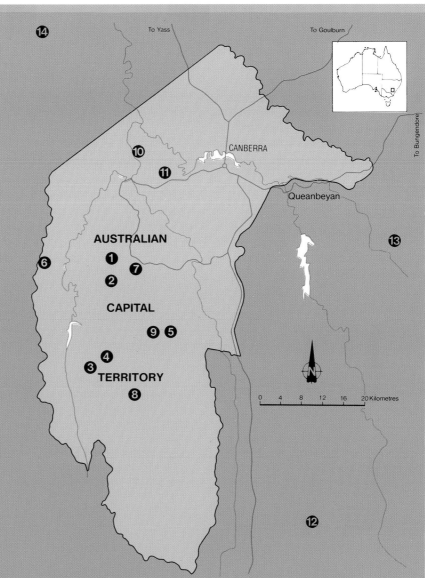

Koalas in and around the Australian Capital Territory

Once common throughout the region where the Australian Capital Territory (ACT) was established in 1927 it seems likely that the only koalas now present in the Territory are descendants of those animals released at Tidbinbilla Nature Reserve in 1939 (1), 1969-71 and 1984 (2), and Orroral Valley (3 & 4) and Bushfold Flats (5) in 1978. Sightings of koalas within the ACT before 1980 (6 & 7) were probably animals which had escaped from Tidbinbilla Nature Reserve. Since 1980 koalas have also been seen near the release sites of 1978 (8 & 9). The history of koalas seen outside the ACT or distant from earlier release points is not clear (10, 11, 12, 13 & 14).

Current Status

Recent records indicate that koalas still occur in and around the Australian Capital Territory. In surrounding New South Wales there have been reports of koalas near Wee Jasper and between Yass and Gunning. To the south they have been seen in the Tinderry Mountains, near Bredbo, and also in the Cooma region.

The current status of free-ranging koalas in the Australian Capital Territory is unclear although there are good reasons to believe that a viable but low density population exists throughout the Tidbinbilla and Brindabella ranges. It is also possible that a remnant race of local koalas is still present in the Territory.

KOALAS IN QUEENSLAND

History

The south-east quarter of Queensland has always sustained the greatest concentrations of koalas, with populations becoming more scattered towards the northern and western boundaries of their range. Queensland has long been considered the stronghold of koalas. As the southern koala populations were being decimated by the fur trade and disease during the late 1800s and early 1900s, koalas apparently remained abundant in the south-east region of Queensland.

As koala hunting was outlawed in the southern States the focus for the fur trade moved to Queensland, at least until 1927 when the final open season of one month took place. The remarkable number of 584 738 koala skins were tendered for sale as a result of this open season. While anecdotal reports suggest that many of these skins were from stockpiles accumulated from illegal hunting during the preceding years, this figure and the estimated 1 000 000 koala furs produced by a six month open season in 1919 testify to the abundance of koalas in Queensland around this time.

Although the final hunting seasons resulted in declines in some koala populations, epidemics were held responsible for a major decline in the number of koalas in the late 1920s and early 1930s. The extent of these population declines was not recorded and it was not until 1967 that any serious attempt was made to survey the koala populations of Queensland.

The 1967 koala survey was organised by the Wildlife Preservation Society of Queensland and was coordinated with a similar survey undertaken by the newly formed New South Wales National Parks and Wildlife Service. Survey forms were distributed throughout Queensland to all primary schools.

A total of 730 koalas were sighted during the survey week of 13 to 19 November 1967. For the full year of 1967, 2650 koalas were reported. From responses to the questionnaire regarding previous knowledge of koalas in the area, Kikkawa and Walter (1968) concluded that there had been some contraction of the koalas'

84. Newspaper report on koala fur sales in Queensland, the *Daily Mail* Brisbane, 15 August 1927.

THE FUR SALES

FALL IN VALUES

KOALA DOWN 10/ A DOZEN

Buyers were not in strong force at the Brisbane fur sales yesterday, and inquiry was limited. Opossum furs were on an average about 5/ a dozen lower than at the sales two weeks ago, and native bear furs 10/ a dozen lower than at the previous auctions.

In the catalogues were about 135.000 opossum furs and about 110.000 native bear furs. Very few of the latter were sold, and substantial withdrawals of opossum furs were made. The lots withdrawn are open for sale by private treaty.

Another sale of furs will be held in Brisbane on September 28, when it is expected that a considerable supply of furs will be offered. The stoppage of the railway services prevented numbers from reaching the market in time to be listed for yesterday's auctions.

84

83. Queensland was the last State to outlaw koala hunting. It is not known how many koalas were killed for the fur trade as during 'closed seasons' trappers probably traded koala pelts as wombat fur.

range in Queensland prior to the survey although there 'was also an indication that recolonisation may even have occurred in some areas in the south-eastern region'.

Ten years later the Wildlife Preservation Society of Queensland repeated the koala survey in Queensland and the population trends suggested by the 1967 survey were confirmed. Questionnaires were once again circulated through the school system and personnel from the Queensland National Parks and Wildlife Service and Forestry Department were asked to assist. The designated survey week was the same as for 1967 during which time 295 koalas were reported. For the year of 1977, Campbell, Prentice and McRae (1979) reported that 1 123 koalas were sighted at 186 sites.

Despite the reduction in the number of koalas observed, which was considered a legacy of the reduced number of questionnaires returned, it was concluded that 'the 1967 and 1977 survey results give no indication of any major contraction or expansion' to the range of koalas in Queensland.

Current status

In Queensland today koalas are widely distributed and locally abundant in a number of areas. Consistent with previous surveys, the greatest concentrations of koalas are in the south-eastern corner of the State. However, there are also some indications that koala populations in northern Queensland may be less abundant than in the past, and that a southward contraction of their range may have taken place.

The national survey of koalas in Queensland was undertaken using a variety of survey forms and newsletters, especially tailored to enhance the response from different parts of the State. While this approach was successful in that a total of 3 100 replies were received, the type of information collected varied greatly and made detailed analysis on a State-wide basis difficult.

In general terms, koalas were positively sighted at 1 122 sites in Queensland during the National

85

85. The south-east quarter of Queensland has always been the stronghold of koalas.

Koala Survey and their presence was inferred from characteristic droppings and scratch marks on trees at a further fifty-four sites. Queensland respondents were also asked to nominate other sites where they knew koalas occurred and 514 such sites were reported. There were 258 reports of koala sightings prior to 1 January 1985, some dating back as far as the 1930s. A total of 1 152 replies registered sites where koalas were not present.

Of the various types of koala survey forms used in Queensland, only those which specifically targeted the rural community requested details of the status of koala populations in surrounding areas. Of 591 replies which recorded recent sightings of koalas, 551 provided some assessment of population status. Sixty respondents (10.9 per cent) suggested that koalas were decreasing in number, sixty-nine (12.5 per cent) said koala populations were increasing, 157 (28.5 per cent) thought koala numbers were stable and 265 (48.1 per cent) were not sure. While there are obvious limitations to this type of subjective assessment, it is perhaps encouraging that overall 41 per cent of those who replied to this part of the survey felt that koala populations were either stable or increasing.

Koalas on Queensland's Islands

Like the islands of Western Port Bay in Victoria, several islands off the coast of Queensland have koala populations. The most southern of these is North Stradbroke Island, in Moreton Bay off Brisbane, where there is a small population of koalas. It is not known whether koalas were introduced onto North Stradbroke Island or they occurred there naturally. It may be significant that none of the islands surrounding North Stradbroke in Moreton Bay have koalas on them.

Further north, koalas are found on four islands off Mackay. The smallest, Newry Island, has a colony of ten to fifteen koalas. Nearby Rabbit Island is home for thirty to thirty-five koalas and the two slightly larger islands of Brampton Island and St Bees Island each have about fifty koalas in residence. There is no information on how these islands were colonised by koalas. The two smallest, Newry Island and Rabbit Island, are joined to the mainland via mangrove areas so it is possible that koalas made their way across this landbridge from the mainland.

Koalas are also present on Magnetic Island, off Townsville. They were probably introduced after mainland clearing activities earlier this century. The number of koalas is repor-

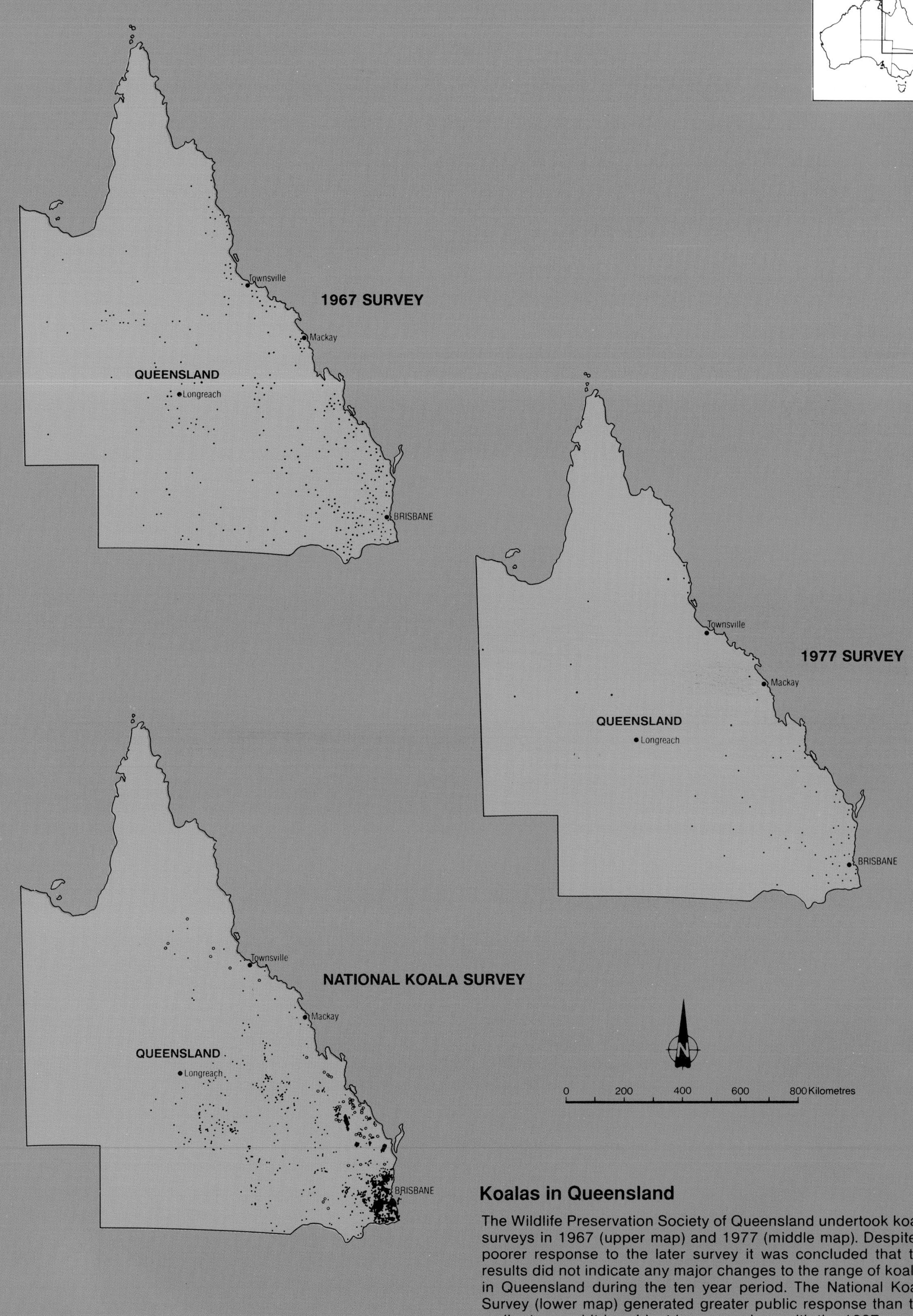

Koalas in Queensland

The Wildlife Preservation Society of Queensland undertook koal surveys in 1967 (upper map) and 1977 (middle map). Despite poorer response to the later survey it was concluded that th results did not indicate any major changes to the range of koala in Queensland during the ten year period. The National Koal Survey (lower map) generated greater public response than th earlier two and it is evident by comparison with the 1967 surve results, in particular, that some inland and northern population may have been lost. Positive sightings (●), inferred presence (o

86. A Queensland koala

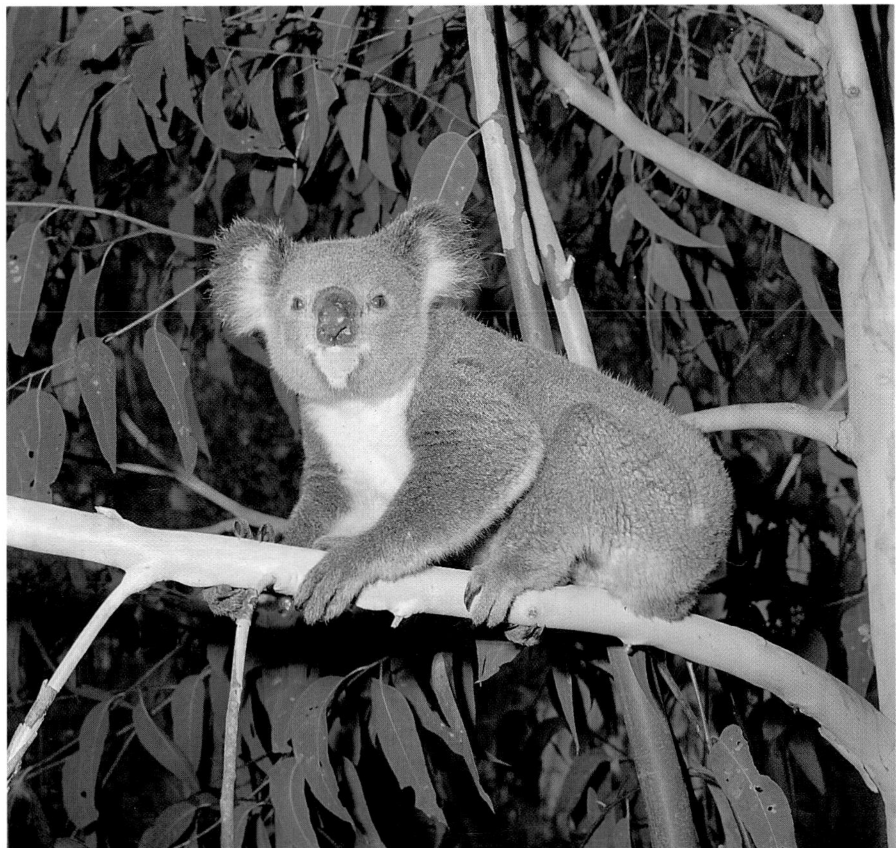

86

ted to fluctuate considerably in this environment. Smith (1987) reported that a cyclone in 1970 significantly affected the koalas and between 1969 and 1977 the population was reduced from 1400 to 300 due to the combined effects of dry weather, fires and dogs.

Unlike their counterparts on the islands of Victoria, overcrowding does not appear to have been a problem for the koala populations on Queensland's islands. Why these populations should show restrained growth is not understood. Diseases resulting from infection with the bacterium *Chlamydia psittaci* cannot be blamed as only isolated cases of conjunctivitis are observed on Brampton Island and Magnetic Island. Long-time koala researcher Frank Carrick of the University of Queensland has treated only a few koalas with 'dirty tail' and 'pink eye' from North Stradbroke Island.

Although koalas are not found there today, anecdotal accounts suggest that Aborigines spoke of koalas on Fraser Island, off Bundaberg, up to the beginning of this century.

A NATIONAL OVERVIEW

Koalas are now widely distributed and locally abundant in many areas of Australia. Stable populations are present throughout much of the suitable habitats of eastern Australia although there is some indication that on the northern and western fringes of their range in New South Wales and Queensland koala numbers have declined in recent years.

The National Koala Survey identified a total of 3 145 sites where koalas were either observed or thought to be present from tell-tale signs. The survey did not set out to estimate the total number of koalas present in the wild. If a species possesses a disparate and fragmentary distribution, population size can sometimes be a misleading indicator of conservation status. In terms of species status and future management options, the 'where are they' question is just as important as the 'how many are there' question.

In the southern States, South Australia and Victoria, the current distribution of koalas has largely been determined by conservation authorities translocating koalas from over-stocked island colonies into suitable mainland areas. These management programs have been so successful that koalas have been re-established throughout large tracts of forest they previously occupied. From being once extinct in South Australia and nearing the same status in Victoria, koalas have now re-colonised these States and populations in many areas continue to grow. With the continuing need to reduce the size of the flourishing French Island and Kangaroo Island 'reservoir' populations through translocations and given appropriate management of koala habitat in these States, the future of koalas in southern Australia should be assured.

At the northern end of their range koalas are most abundant in the north-east corner of New South Wales and the south-east corner of Queensland. Both areas have rapidly expanding urban centres likely to threaten habitat occupied by koalas. Unless land management practices takes account of the habitat needs of koalas then local extinctions are inevitable. On a larger scale, and more optimistic note, there are a number of stable or enlarging koala populations within this area, which occupy more secure habitats. Assuming these populations remain viable, the future of koalas in northern Australia should also be secure. The remarkable recovery of koalas in southern Australia over the past seventy years clearly demonstrates how readily koala populations can be managed, if necessary, to augment dwindling populations or recolonise areas which once supported koalas.

While the widespread distribution of koalas and their apparent capacity to recover from very low population numbers is reassuring, it is evident that in modern Australia they face an ever increasing number of threats. It is only through recognition of these threats that appropriate action can be taken to ensure that koalas are not, once again, reduced to low numbers and their survival threatened.

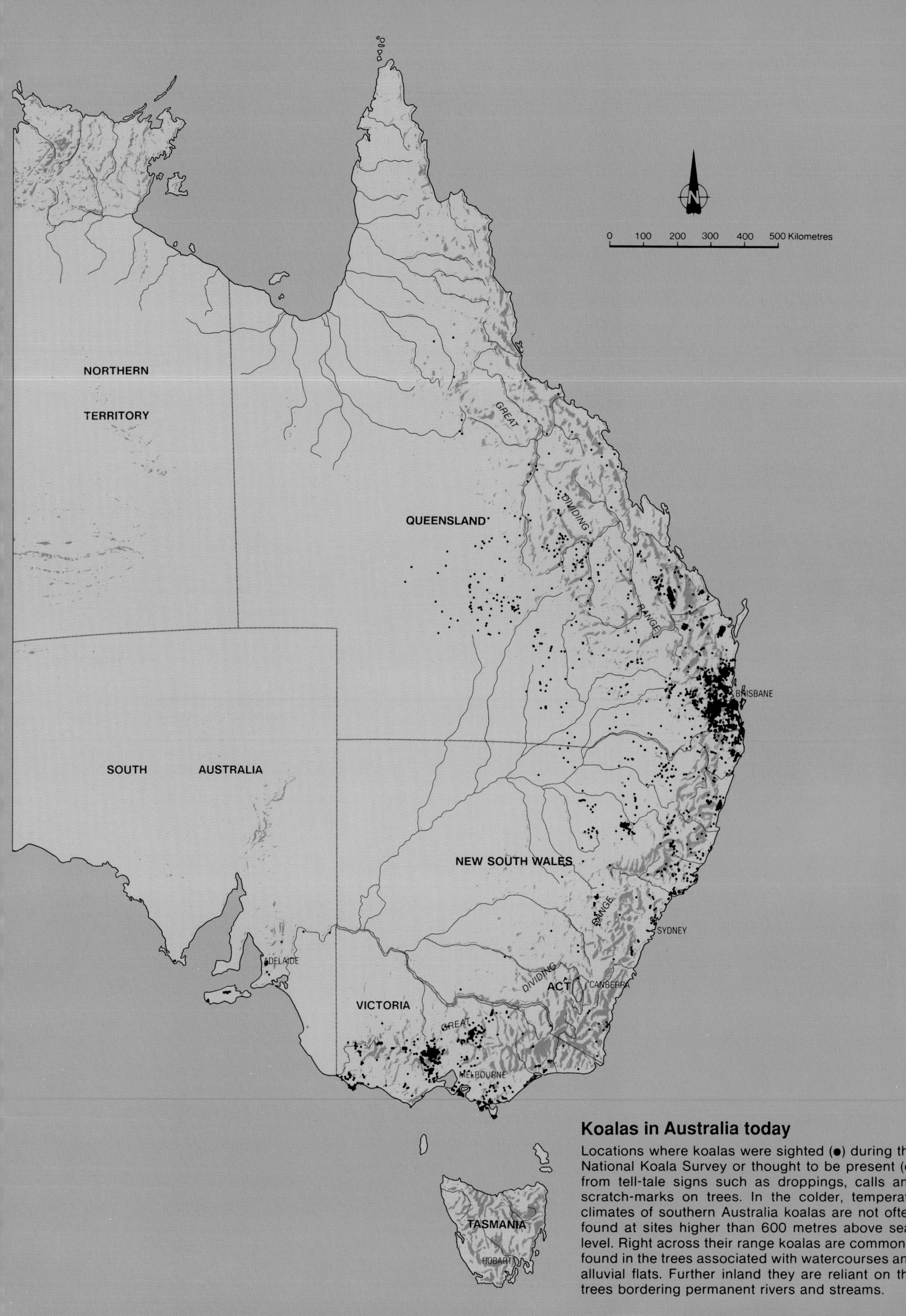

Koalas in Australia today

Locations where koalas were sighted (●) during th National Koala Survey or thought to be present (from tell-tale signs such as droppings, calls an scratch-marks on trees. In the colder, tempera climates of southern Australia koalas are not ofte found at sites higher than 600 metres above sea level. Right across their range koalas are common found in the trees associated with watercourses an alluvial flats. Further inland they are reliant on th trees bordering permanent rivers and streams.

WHY WE FIND KOALAS WHERE WE DO

Describing the ideal habitat for koalas is not a simple task. Their range in Australia spans nearly twenty-two degrees of latitude, from the steamy tropics of northern Queensland through to the cool temperate climate of southern Victoria. Explaining why koalas occur where they do is made difficult because of the range of different ecosystems they occupy and the uneven nature of their distribution across this extensive range. As early as 1863, noted naturalist John Gould made this same observation and remarked that the koala was 'extremely local in its habitat'. It seems, therefore, that the patchy distribution of koalas is not a recent phenomenon which can be dismissed as a by-product of European settlement in Australia.

Although our understanding of the habitat needs of koalas has improved greatly since Gould's time, the patchiness in their distribution is not yet fully understood. Koalas are curiously absent from large areas of seemingly suitable forest. While this may be partly a legacy of past hunting and devastating epidemics, much remains to be learnt about why koalas prefer some habitats over adjacent areas apparently offering similar environmental and physical conditions. If the future of koalas is to be assured, the habitat needs of the species must be described in detail.

In looking at the distribution of koalas at both the regional and local levels, it becomes evident that the habitat requirements of koalas differ across their range. Koalas in Victoria have very different habitat needs to those in Queensland and as a consequence the measures taken to protect koalas must be tailored to the particular environmental situation.

Dietary Preferences

The ability to survive on a diet of eucalypt foliage sets koalas apart from all other mammals, except the Greater Glider (*Petauroides volans*). This dietary specialisation fundamentally determines the distribution of these species.

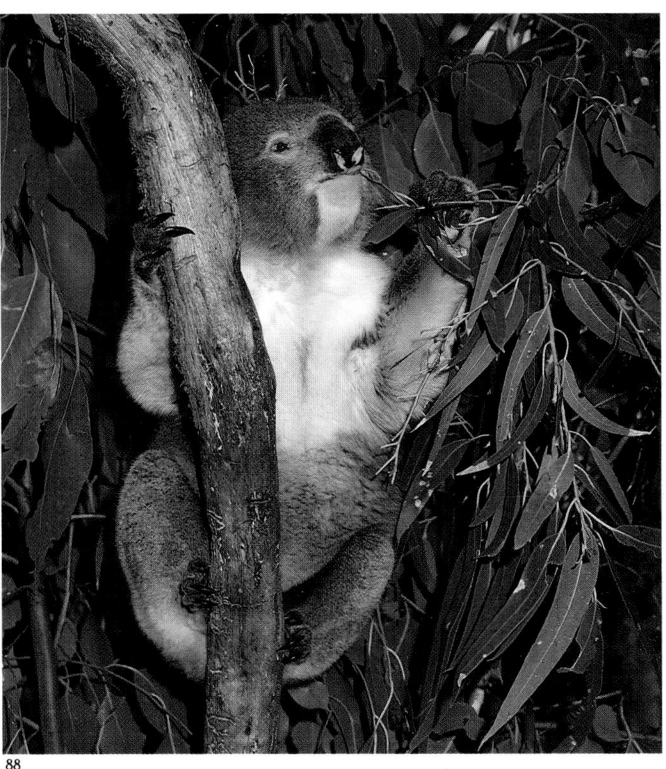

88

DURING my two years' ramble in Australia, a portion of my time and attention was directed to the fauna of the dense and luxuriant brushes which stretch along the south-eastern coast, from Illawarra to Moreton Bay. I also spent some time among the cedar brushes of the mountain ranges of the interior, particularly those bordering the well-known Liverpool Plains. In all these localities the Koala is to be found, and although nowhere very abundant, a pair, with sometimes the addition of a single young one, may, if diligently sought for, be procured in every forest. It is very recluse in its habits, and, without the aid of the natives, its presence among the thick foliage of the great *Eucalypti* can rarely be detected. During the daytime it is so slothful that it is very difficult to arouse and make it quit its resting-place. Those that fell to my own gun were most tenacious of life, clinging to the branches until the last spark had fled. However difficult it may be for the European to discover them in their shady retreats, the quick and practised eye of the aborigine readily detects them, and they speedily fall victims to the heavy and powerful clubs which are hurled at them with the utmost precision. These children of nature eat its flesh, after cooking it in the same manner as they do that of the Opossum and the other brush animals.

I believe the Koala to be extremely local in its habitat, as up to the present time the south-eastern portion of the continent of Australia is the only part in which it is known to exist.

87

87. Excerpt from *The Mammals of Australia* by John Gould, 1863.

88. Koalas are not nearly as fussy about food as was once thought. During the National Koala Survey koalas were seen to be eating or sitting on sixty-nine different eucalypts and about thirty non-eucalypt tree types.

89. Koalas sometimes use non-eucalypt trees, like this *Grevillea*, as refuge from extreme heat or domestic dogs. What is surprising is that they have been known to eat the leaves too.

The earliest attempts at keeping koalas in captivity failed and they gained a reputation as fussy eaters with a penchant for only a few types of eucalypt leaves. Following an unsuccessful attempt to transport live koalas to England, Barrallier noted in the *Sydney Gazette* of 1803 that a koala's 'food consists solely of gum leaves in the choice of which it is excessively nice (*fastidious*)'. Barrallier was wrong on both counts although his view of koala eating habits persisted until just a few years ago, largely sustained by observations of the relatively limited feeding behaviour of captive animals.

Over the past fifteen years several studies have shown that koalas, although showing preference for the leaves of a few key eucalypt species, eat a wide range of others as well as the leaves of non-eucalypt trees. In a recent study, Murray (1988) collated information about which trees koalas preferred from twenty-one sources and found that, Australia-wide, they have been observed eating or sitting on 120 different eucalypt and thirteen non-eucalypt tree species. This is a good indication of the diverse dietary habits of koalas, as normally they are found sitting on the trees they fed from during the previous night or will feed from during the forthcoming evening.

Details of the tree species occupied by koalas were also collected as part of the National Koala Survey and reinforce the view that koalas are more generalist in their eating habits than once thought. During the national survey, koalas were observed eating or sitting in sixty-nine eucalypt species and almost thirty different non-eucalypt tree types. The non-eucalypt species included a range of native plants such as wattles, tea trees and paperbarks as well as introduced species such as the Monterey Pine (*Pinus radiata*).

Describing the ideal habitat of koalas is further complicated by regional differences in their staple eucalypt food trees. In southern Australia the major koala food tree is the Manna Gum (*Eucalyptus viminalis*) whereas in Queensland the Forest Red Gum (*Eucalyptus tereticornis*) represents most of the koala's diet. Another ten to fifteen eucalypts are routinely listed as popular with koalas although there is little agreement among researchers as to what comprises the most preferred food trees.

The State-by-State results of the national survey provide a good illustration of how the dietary preferences of koalas vary across their range.

In South Australia, 73.3 per cent (148) of 202 reports were of koalas consuming or occupying Manna Gum, 11.9 per cent (24) were observed on South Australian Blue Gum (*Eucalyptus leucoxylon*) and 5.4 per cent (11) were found on River Red Gum (*Eucalyptus camaldulensis*). The remaining observations of koalas in trees were divided between a further seven eucalypt species with one koala seen on a Monterey Pine (*Pinus radiata*) and another on a Cypress Pine (*Callitris sp.*)

During the National Koala Survey in Victoria there were 983 records of koalas identified with tree species. Although some reports listed the three or four most common eucalypts in the surrounding area, Manna Gum was still the most popular with nearly 50 per cent of koalas recorded in this species. The other main tree species were Messmate Stringybark (*Eucalyptus obliqua*), with 8.2 per cent of observations, followed closely by Swamp Gum (*Eucalyptus ovata*), Narrow-leaved Peppermint (*Eucalyptus radiata*), Tasmanian Blue Gum (*Eucalyptus globulus*), and River Red Gum. Victorian koalas were observed on a total of twenty-eight eucalypt species and ten non-eucalypts with seventeen further reports indicating koalas occupying *Acacia* (nine), *Banksia* (seven) and *Casuarina* (one) species.

In New South Wales only Forestry Commission personnel were asked to provide details of the trees occupied by koalas. In total, koalas were observed on twenty-five eucalypt and four non-eucalypt tree types. Most popular with New South Wales koalas were the Sydney Blue Gum (*Eucalyptus saligna*), followed by Tallow-wood (*Eucalyptus microcorys*),

89

Grey Gum (*Eucalyptus punctata*), Blackbutt (*Eucalyptus pilularis*), Flooded Gum (*Eucalyptus grandis*) and Small-fruited Grey Gum (*Eucalyptus propinqua*). Koalas were also found on Black Wattle (*Acacia mearnsii*), Blackwood (*Acacia melanoxylon*), Forest Oak (*Allocasuarina torulosa*) and Brush Box (*Lophostemon confertus*). In their report of the National Koala Survey in New South Wales, Reed, Lunney and Walker (1988) point out that, from this and earlier sources, koalas have now been recorded on fifty-five eucalypt and eleven non-eucalypt tree species in that State.

In the National Koala Survey in Queensland information of the trees species occupied by koalas was given in 1 063 reports. Thirty-five eucalypt and eighteen non-eucalypt species were reported as hosting koalas. Just over 40 per cent of the 1 018 eucalypt-related sightings were of koalas in Forest Red Gum. Next most common were the Narrow-leaved Red Ironbark (*Eucalyptus crebra*), at 7.5 per cent followed by Poplar Box (*Eucalyptus populnea*), River Red Gum, Spotted Gum (*Eucalyptus maculata*), and Small-fruited Grey Gum. Of the non-eucalypts, koalas were most frequently observed in Five-veined Paperbark (*Melaleuca quinquenervia*), Brigalow (*Acacia harpophylla*) and Brush Box (*Tristania conferta*).

Tree species preferred by koalas

Across their range koalas show considerable differences in food preference as shown by the results of the National Koala Survey presented here. In total, koalas were seen sitting in or eating sixty-nine eucalypt species and nearly thirty different non-eucalypt tree types.

90 & 91. The Monash University research team has used radiotransmitters fitted to collars of koalas to follow their exploits after they have been released at new sites. This is part of the ongoing translocation program in Victoria. Kath Handasyde holds the radiotracking antenna.

While the Australia-wide dietary information indicates how adaptable koalas are in terms of selecting their food trees, it seems that trees of several key eucalypt species must be present in an area in order to sustain a viable koala population.

Much of the disagreement over what constitutes the primary food trees of koalas has resulted from local and individual variations in feeding habits. These variations may be the result of social or seasonal factors or, alternatively, may reflect a change in dietary preferences forced upon koalas by the food trees available at any particular time or location.

Locally, dietary habits are closely tied with the social spacing of koalas and the tree species present within the home range of each individual animal. A number of researchers have noted koalas within localised populations showing preferences for different eucalypt species. Hindell (1984) observed that koalas seem to consume different species of eucalypts in the proportion that they are available within their home range.

In Victoria, Lee and Martin (1988) found that koalas almost invariably disperse from the point of release when taken from over-crowded sites and relocated to other localities with the same dominant eucalypt species. By radio-tracking individual koalas, they discovered that relocated koalas establish home ranges in a variety of habitats, sometimes necessitating a change in dietary habits to suit the new home. The solitary nature of koalas appears to play a key role in determining dietary habits with the desire for social segregation, on occasions, overriding previously expressed food preferences.

Koala food preferences also appear to alter with the seasons. In the Brisbane Ranges, near Melbourne, Hindell (1984) found that during summer koalas showed equal preference for Manna Gum and Swamp Gum whereas Manna Gum dominated their diet in all other seasons. Whether the palatability of eucalypts alters seasonally or some other factor is involved is not known.

90

A number of researchers have found that koalas having a preference for a particular species of eucalypt also show preference for individual trees. Why koalas select individual trees over others of the same species is not known although a clear preference for bigger trees has been noted. Under certain circumstances eucalypts are believed to produce so-called 'defence compounds', designed to discourage insects and other foliage eaters. Whether 'defence compounds' make some trees less flavoursome to koalas and therefore alter their feeding habits is currently under investigation.

At the individual level, young koalas probably develop a 'taste' for the same eucalypts as their mother while riding on her back during the weaning period. Once independent, young koalas usually remain in the home range of their mother for another twelve months, further reinforcing their dietary habits. Young female koalas then establish a home range of their own, usually nearby or overlapping that of their mother. The eucalypts they encounter when young probably become their staple diet thereafter.

In contrast, young male koalas are evicted from the maternal home range when about two years old. They are then nomadic for up to three years or until such time as they attain sufficient body size to establish a home range of their own. While moving around, the young males are probably forced to occupy sub-optimal habitats and may therefore be forced to adopt a more generalist attitude to food. Hindell (1984) sug-

91

92. Young koalas probably develop a 'taste' for the same eucalypts as their mother while riding on her back during the weaning phase.

92

gested that a group of male koalas he observed showing preference for Red Stringybark (*Eucalyptus macrorhyncha*), rather than the Manna Gum preferred by the rest of the Brisbane Ranges' population, were in fact young males denied access to the prime habitat areas.

As researchers have begun to look closely at the eating habits of koalas more and more reports of them sitting on or eating non-eucalypt trees have been received. During the national survey koalas were observed occupying nearly thirty different non-eucalypt tree species. Although some of these observations can be dismissed as koalas seeking temporary refuge from extreme heat or dogs, it appears koalas can eat leaves from a wide range of non-eucalypt trees.

Koalas have been observed eating a number of different wattle, tea tree and paperbark species and Lee and Martin (1988) reported that koala droppings found beneath Monterey Pine trees on French Island contained pine needles. Koalas have also been observed in pine forests at several other locations. Nevertheless, it is clear that koalas only occasionally dine on non-eucalypt foliage. It is not known under what circumstances they abandon their normal diet nor how nutritious these alternative foods are to koalas.

Climatic considerations

Factors other than the availability of food trees also affect the distribution of koalas. Once again the perceptive writings of John Gould (1863) provide some of the earliest comments on the habitat preferences of koalas. Gould wrote that koalas were to be found among 'the dense and luxuriant brushes which stretch along the south-eastern coast'. Despite his relatively restricted observations, Gould accurately noted the higher concentration of koalas in the coastal forests, an observation supported by the present distribution of the species.

Koalas occur at their greatest concentrations in coastal lowlands and become less common with increasing height above sea level and decreasing annual rainfall. Lee and Martin (1988) maintain that in southern Australia koalas are rare or absent in wet forests above 600 metres elevation. This is supported by the findings of the National Koala Survey in Victoria. Of the 696 localities where koalas were positively sighted in Victoria, 584 of these (83.9 per cent) were at elevations less than 600 metres. No koalas were seen at sites above 1 000 metres with only three localities recorded above 800 metres. In contrast, koalas were reported from 292 sites (42.0 per cent) within 200 metres of sea level.

While the current distribution of koalas in Victoria has largely been determined by wildlife authorities the absence of koalas from chilly Tasmania lends some support to the view that climatic factors restrict koalas from some areas of otherwise suitable forests in the highlands of southern Australia.

93

94

93. Leaves are inserted into the mouth at an oblique angle.

94. A koala drinking from a stream.

95. The food and digestive system of the koala.

THE DIGESTIVE SYSTEM OF KOALAS — COPING WITH A DIET OF EUCALYPT LEAVES

Koalas spend about 20 per cent of their lives eating, almost 80 per cent sleeping and resting and less than 1 per cent moving from tree to tree and seeking mates. This relatively lethargic lifestyle is indicative of the physiological adaptations enabling them to live in the trees and survive on a diet consisting largely of the leaves and buds of eucalypts.

Through inactivity and slow metabolism, koalas survive on a diet estimated to yield only half the energy needed by other similar sized mammals. Nagy and Martin (1985) found that during winter koalas need to consume just over 500 grams of eucalypt leaves daily in order to meet their energy demands. In our terms, this is equivalent to the energy we get from about 200 grams of breakfast cereal.

The adaptation of koalas to a leaf diet begins with the selection process. Young leaves are preferred but as these are not available throughout the year great care is taken in selecting mature leaves. The koala's acute sense of smell is used to recognise the more palatable foliage.

Koalas break up eucalypt leaves initially by using their specialised teeth in a grinding, chewing action. They insert single leaves into their mouths at an oblique angle, via the gap between the incisors and the premolars. The premolars sever the stem connection and the molars grind and pulverise the leaf to allow the enzymes in the saliva to begin the breakdown processes. To hasten the mastication phase, the leaf is swivelled from one side of the mouth to the other so that the grinding molars attack the leaf at different angles.

Passing into the relatively small stomach, the food is attacked by acids and enzymes which continue the digestive process by beginning the breakdown of sugars and proteins. The stomach is where most of the toxic compounds found in eucalypt leaves are liberated and absorbed into the blood. These compounds are subsequently filtered from the blood by the liver, modified to remove their toxicity and then excreted as bile or urine. After extended exposure to air, the phenol in koala urine darkens, staining the tree or ground where it falls.

Koalas have few competitors, other than insects, for the foliage of eucalypt trees. Only a few animal species are able to cope with the levels of toxic substances, essential oils and phenol contained in eucalypt leaves. To most other animals, including humans, these compounds can be fatal. Essential oils give eucalypt forests their characteristic fragrance and phenolic compounds are contained in the tannins which may comprise as much as 25–30 per cent of the favourite food of koalas.

As the food moves into the koala's small intestine further breakdown of proteins takes place. Fats are digested and the accumulated nutrients absorbed into the bloodstream.

Further down the digestive tract, the food moves into part of the hindgut called the caecum; a greatly enlarged cul-de-sac of the large intestine corresponding to the human appendix. The caecum, which is up to 2 metres long and 10 centimetres in diameter, houses micro-organisms capable of breaking down the carbohydrates in eucalypt leaves.

Eucalypt foliage contains 30–50 per cent carbohydrates in the form of fibre (cellulose and lignin) which is bulky and produces relatively little energy. Like all herbivorous mammals, koalas lack the digestive enzymes needed to breakdown cellulose, relying instead upon micro-organisms to gain energy from carbohydrates. The micro-organisms survive by breaking down the carbohydrates via a fermentation process. Any excesses in the energy-yielding products of this process are made available to the koala. Cork (1987) has estimated that koalas gain about 10 per cent of their energy by courtesy of the micro-organisms in the hindgut. Contrary to common opinion, the fermentation process which

What's in a eucalypt leaf ?

fat 8 %
sugars/starch 5 %
minerals 2 %
fibre 18%
protein 4 %
water 50 %
tannins 13

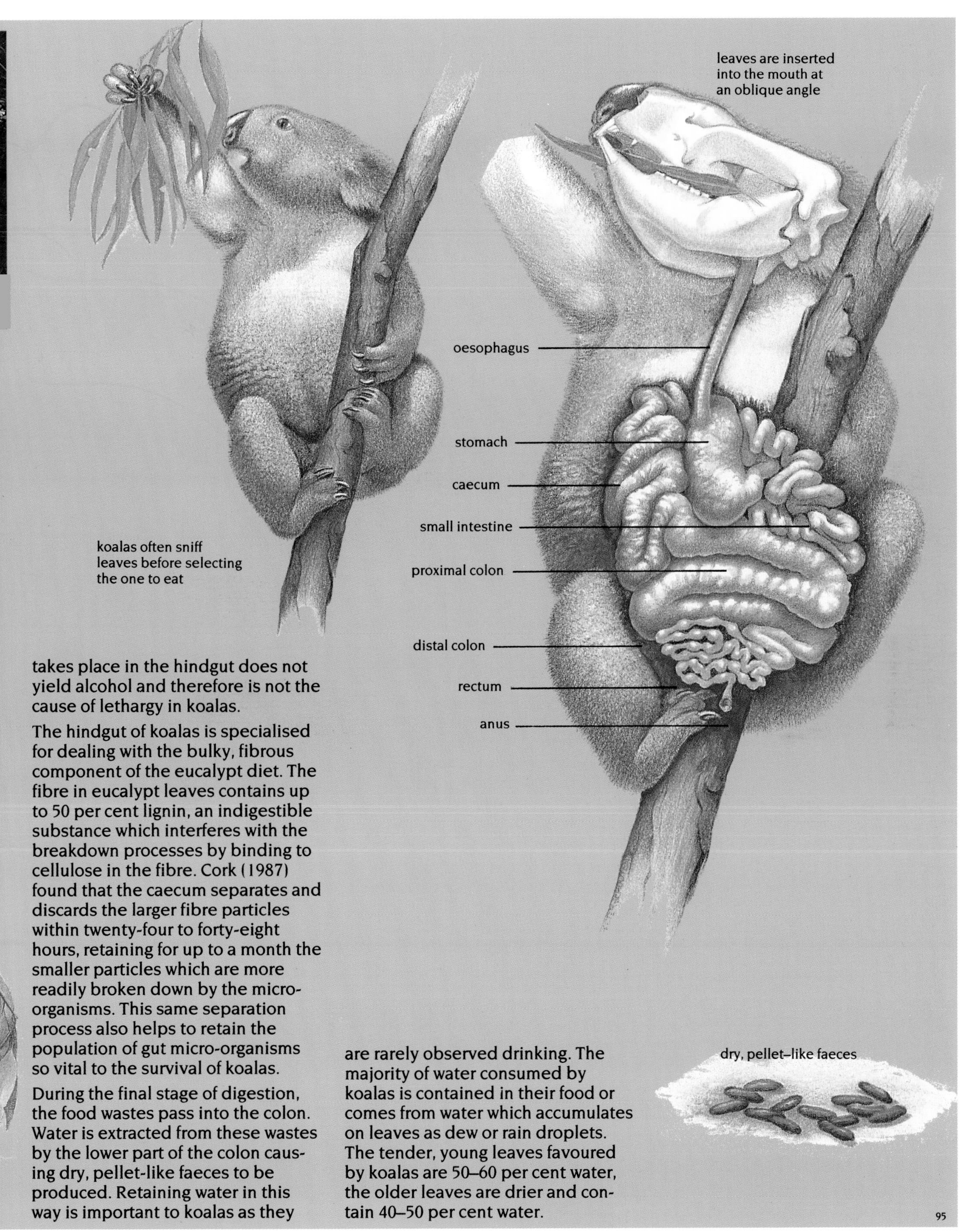

koalas often sniff leaves before selecting the one to eat

takes place in the hindgut does not yield alcohol and therefore is not the cause of lethargy in koalas.

The hindgut of koalas is specialised for dealing with the bulky, fibrous component of the eucalypt diet. The fibre in eucalypt leaves contains up to 50 per cent lignin, an indigestible substance which interferes with the breakdown processes by binding to cellulose in the fibre. Cork (1987) found that the caecum separates and discards the larger fibre particles within twenty-four to forty-eight hours, retaining for up to a month the smaller particles which are more readily broken down by the micro-organisms. This same separation process also helps to retain the population of gut micro-organisms so vital to the survival of koalas.

During the final stage of digestion, the food wastes pass into the colon. Water is extracted from these wastes by the lower part of the colon causing dry, pellet-like faeces to be produced. Retaining water in this way is important to koalas as they are rarely observed drinking. The majority of water consumed by koalas is contained in their food or comes from water which accumulates on leaves as dew or rain droplets. The tender, young leaves favoured by koalas are 50–60 per cent water, the older leaves are drier and contain 40–50 per cent water.

dry, pellet–like faeces

In Queensland, the distribution of koalas extends as far north as that of their favoured food tree the Forest Red Gum. Koalas are generally found in fewer numbers on the western side of the Great Dividing Range although some large populations are found within remnant pockets of suitable forest habitat. Towards the western margin of their range koalas become more scattered and are almost invariably found in the trees bordering permanent watercourses. Smith (1987) has suggested that the 'line of 50 centimetres annual rainfall appears, with few exceptions, to be the limit' to the westward range of koalas.

Do Koalas Care About the Structure of the Forest They Live in?

It has long been recognised that koalas are less common in forests where the tree canopy is continuous or closed. The results of the National Koala Survey support this as they show a clear preference by koalas for the more open forests and woodlands.

In South Australia and Victoria observers recorded the forest type details at a total of 533 sites. Of these, 44.6 per cent (238 sites) were recorded as 'woodland' and 33.4 per cent (178 sites) as 'open forest'. Only twelve or 2.3 per cent of sites where koalas were found were classied as 'closed forest'.

In New South Wales, participants in the national survey were not asked to provide information on forest types. However, the results of the survey in Queensland showed that koalas from northern Australia have the same preference for more open forests as their southern counterparts. Of 528 mostly rural sites where koalas were seen and forest type information provided, 7.4 per cent (thirty-nine sites) were listed as 'woodland', 90.0 per cent (480 sites) as 'open forest', and only six sites or 1.1 per cent were classed as 'closed forest'. Even allowing for the greater difficulty of observing koalas in closed forest, the evidence suggests koalas prefer the more open forests and woodlands.

Proximity to Water?

During the National Koala Survey many reports described koalas living on the trees along gullies and alluvial flats bordering watercourses. While this is not surprising for the more arid parts of their range, such as inland Queensland, it is interesting that the pattern persisted in the wetter and more southern areas.

In areas of intensive farming activity, the close association of koalas with rivers and streams largely reflects past land clearing practices where the trees immediately bordering watercourses were routinely left standing. This retention of habitat corridors along streams and rivers may have unwittingly been the saviour of koalas in many areas which were subjected to large-scale forest clearing.

Several studies in native forests have now demonstrated that koalas, and other tree-dwelling marsupials, tend to be concentrated in these areas. In their analysis of the results of the National Koala Survey in New South Wales, Reed, Lunney and Walker (1988) have shown that koala distribution, in that State at least, is closely correlated with those species of eucalypt growing on relatively nutrient rich soils. In the Brisbane Ranges of Victoria, Hindell (1984) found that koalas preferred forests dominated by Manna Gum and Swamp Gum — species mostly found on the more fertile and deeper soils associated with watercourses.

In summary, koalas prefer open forest and woodland in areas without extremely cold weather. They appear to favour the foliage of eucalypts growing in the deeper, more fertile soils of gullies and alluvial flats and are therefore frequently found in trees close to rivers and streams. Koalas also show a preference for larger rather than smaller trees. In southern Australia their staple food tree is Manna Gum while northern populations prefer Forest Red Gum.

KOALA 'BISCUITS'

Koalas are renowned for their fussy eating habits. This was recognised after the first attempts to keep koalas in captivity failed. The frustrations experienced by the early naturalists and scientists are still felt today even though there is a better understanding of koala dietary needs and feeding habits.

Laboratory-based research on koalas has always been hampered by their dietary requirements. Koalas consume up to half a kilogram of eucalypt leaves daily which means their keepers must devote a great deal of their time and financial resources to collecting suitable fresh food.

In 1985 Pahl and Hume, at the University of New England in Armidale, made the first attempts to produce a koala 'biscuit'. The idea was to produce an artificial food that imitated the nutritional characteristics of the eucalypt leaf so as to relieve the burden of regular forays into the forest to collect food.

The first task was to decide what to put in the biscuits so that they would be nutritionally suitable and flavoursome to koalas. In order to determine what leaves koalas preferred, 'cafeteria-style' experiments were conducted to rank eighteen local eucalypt species.

After determining the most preferred foliage, Pahl and Hume carried out a detailed analysis of the leaves to isolate the constituent parts and their relative proportions. From this information, a recipe for koala biscuits was derived.

To encourage the koalas to eat the biscuits, and make them easy to manipulate and chew, the biscuits were made pliable, 2 millimetres thick, 15 millimetres wide and 60 millimetres long. At first, the koalas showed no interest at all in the artificial diet, possibly not realising that the strange looking biscuits were edible. Now, several years later and after a number of modifications to the biscuits, Pahl and Hume have been greatly encouraged by the results of their feeding trials.

Weaning koalas onto an artificial diet is a long and painstaking experience requiring many hours of hand-feeding. The biscuits are dipped in a vitamin-enriched paste of ground eucalypt foliage and a little water before they are given to the koalas to eat. As yet, no koalas have been completely weaned onto the artificial diet. However, by providing koalas with the biscuits and paste the intake of natural foliage has been reduced by about 40 per cent. This result is most encouraging as it indicates that artificial food can be used to supplement the natural diet. Eucalypt biscuits could one day completely replace natural foliage for captive koalas.

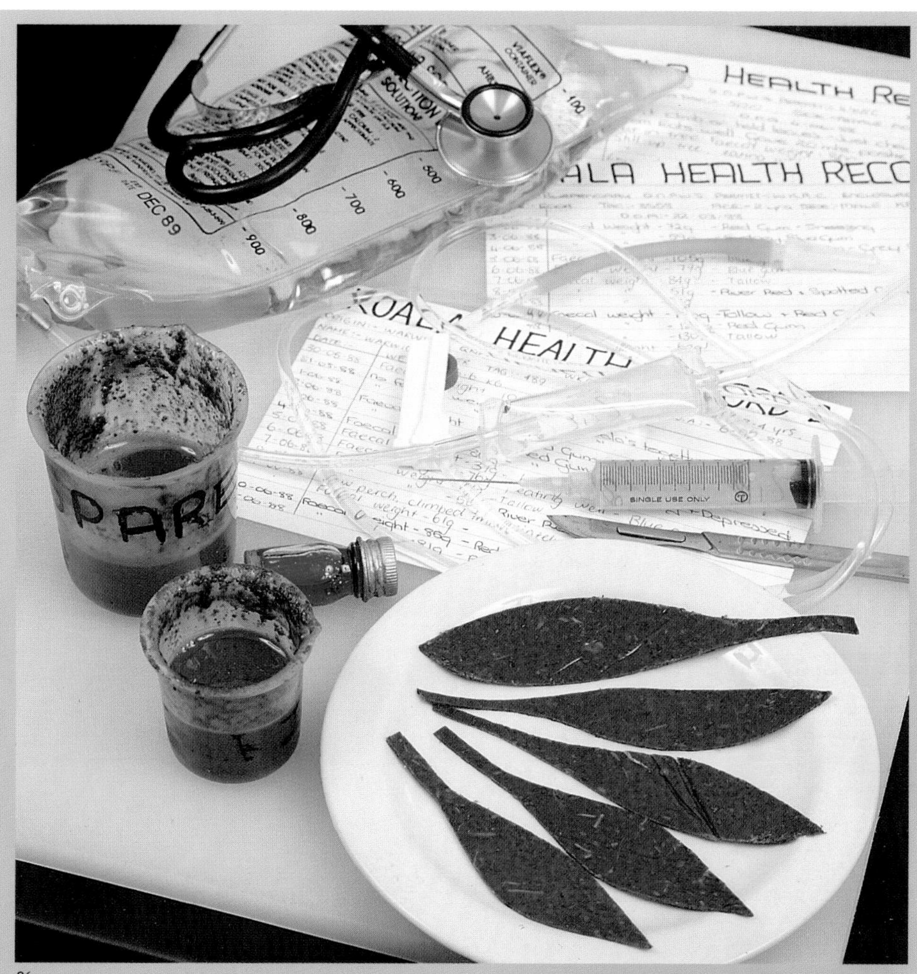

96

96. The 'biscuits' are made pliable and leaf-shaped and are dipped in vitamin-enriched eucalypt paste to encourage the koalas to eat them.

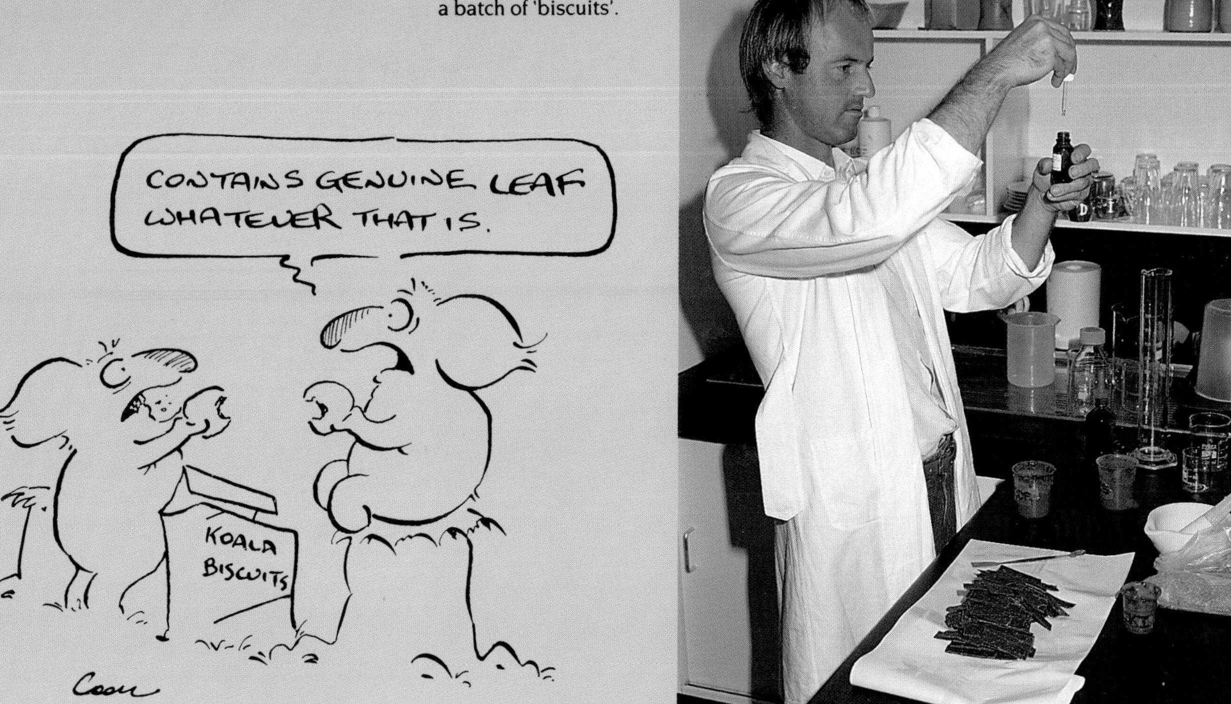

CONTAINS GENUINE LEAF WHATEVER THAT IS.
KOALA BISCUITS

97

97. Lester Pahl carefully measures out the additives for a batch of 'biscuits'.

THE HUMAN PREDATOR

98. Ladysmith, New South Wales, 1910

99. In the Yass district — 'An Australian Hayfield', *c.* 1910.

100. Humans are the greatest threat to koalas today

98

99

100

Vast areas of Australia were covered with eucalypt forest when European settlers first arrived in 1788. There seems little doubt that the forests of eastern Australia supported flourishing populations of koalas extending from north Queensland through to South Australia. Although the Aborigines took their toll on koalas and dingoes and bushfires affected their numbers in some areas koalas were, in general, plentiful across their range.

Soon after establishing the first colonies along the coast the early settlers began to explore the rugged inland areas. The more intrepid amongst them claimed and worked large pieces of land, their 'selections'. Villages sprang up to service the increasing number of pioneers working the land and a rudimentary road system was quickly developed. As the push inland continued the clearing of land for farming accelerated.

The first farmers sought the alluvial flats by rivers and streams where the soil was deep and fertile. These were also the forest areas where koalas lived in greatest numbers. As the first trees were felled to make way for farms and villages, massive soil erosion occurred and resulted in downstream pollution. In 1788 Governor Phillip imposed the first tree preservation order to protect the Tank Stream on which the growing settlement, ultimately to become Sydney, so depended. A new era had dawned in the Great Southern Land and koalas were to be just one species to suffer.

Perhaps ironically, in some areas European settlement initially benefitted koalas as Aborigines moved away and the early farmers killed the dingoes. However, respite from these predators was short-lived as in the wake of the Aboriginal hunters came the white hunters with their firearms and poisons. The trade in koala furs grew rapidly and koala hunting became widespread, serving to supplement the income of settlers attempting to eke an existence from the land. Uncontrolled koala hunting resulted in rapid declines in the koala populations of South Australia,

Victoria and New South Wales, before it was eventually outlawed after an open season in Queensland in 1927 yielded nearly 600 000 koala furs in one month.

The Loss of Koala Habitat

As Australia developed into a nation whose economy focussed on primary production, an increasing number of native forests were cleared. With population growth, more land was also dedicated to urban development and it is significant that the three largest urban centres, Melbourne, Sydney and Brisbane are founded on what were once prime koala habitats.

When the first European settlers arrived in Australia the total area of tall to medium-sized trees in the four eastern States where koalas lived is estimated to have been just over 1 230 000 square kilometres. In their 1984 report, Wells, Woods and Laut, from the Commonwealth Scientific and Industrial Research Organisation, suggested that just over half of these forests (nearly 670 000 square kilometres) have been removed or severely modified.

In just 200 years, forest removal has meant that koalas are no longer present in extensive areas where they once were and that their future in some areas can no longer be assured.

Koalas in Rural Areas

In parts of Australia used for primary production extensive areas of land have been cleared in order to maximise the areas available for the cultivation of grain crops, grazing of livestock, production of dairy produce, market gardening, fruit growing and so on. In many areas strips of eucalypt forest have been retained as a defence against soil erosion, as windbreaks, as shade trees along rivers and streams or as part of the roadside verge. These strips of native vegetation also serve as corridors linking larger areas of natural forest. Forest remnants are frequently on rocky ridge tops where the soils are shallower and less fertile. The eucalypts that grow in these areas are not those koalas prefer and

101

102

103

101,102 & 103. Further west the larger-scale clearing of land for grain crop cultivation and livestock grazing has left only ribbons of trees along rivers and roads. Such corridors sometimes link forest remnants although these are commonly on rocky ridges or hilltops where the soils are less fertile and where the trees are not those preferred by koalas.

104

104. Intensive farming activities along the coastal hinterland of eastern Australia have removed extensive areas of once prime koala habitat.

105

108

106

105. Many koalas in rural areas now find themselves in habitat islands surrounded by grazing land.

106 & 107. Satellite suburbs bring with them many hazards for koalas and require large-scale clearing so that roads and essential services are available.

108. The continued expansion of urban centres into areas of native forest poses the greatest threat to koalas.

107

therefore, are less than ideal for their needs.

While these corridors snaking along watercourses and beside roads permit some movement between koala populations, increasingly koala communities in rural areas are being isolated within small forest 'islands'.

Our Koala Unfriendly Society

The coastal hinterland to the east of the Great Dividing Range has always been the stronghold of the northern koala populations. It is here that land clearing has had, and will continue to have the most significant impact on koalas. As tourist and housing developments continue to expand along the eastern seaboard, more and more koala habitats will inevitably be lost. Unfortunately, areas of prime real estate value are often also areas where large numbers of koalas are found.

With urban expansion, satellite suburbs are pushing further and further into koala habitats and great swathes of forest are cleared to make way for road and rail links and essential services such as electricity and water. Although land clearing to accommodate urban expansion has not removed as much forest as rural practices in the past, the effect has been the same — fragmentation of koala habitats.

In general, koalas and modern society do not mix. People tend not to give consideration to the needs of koalas, and consequently, koalas cannot survive in towns and cities.

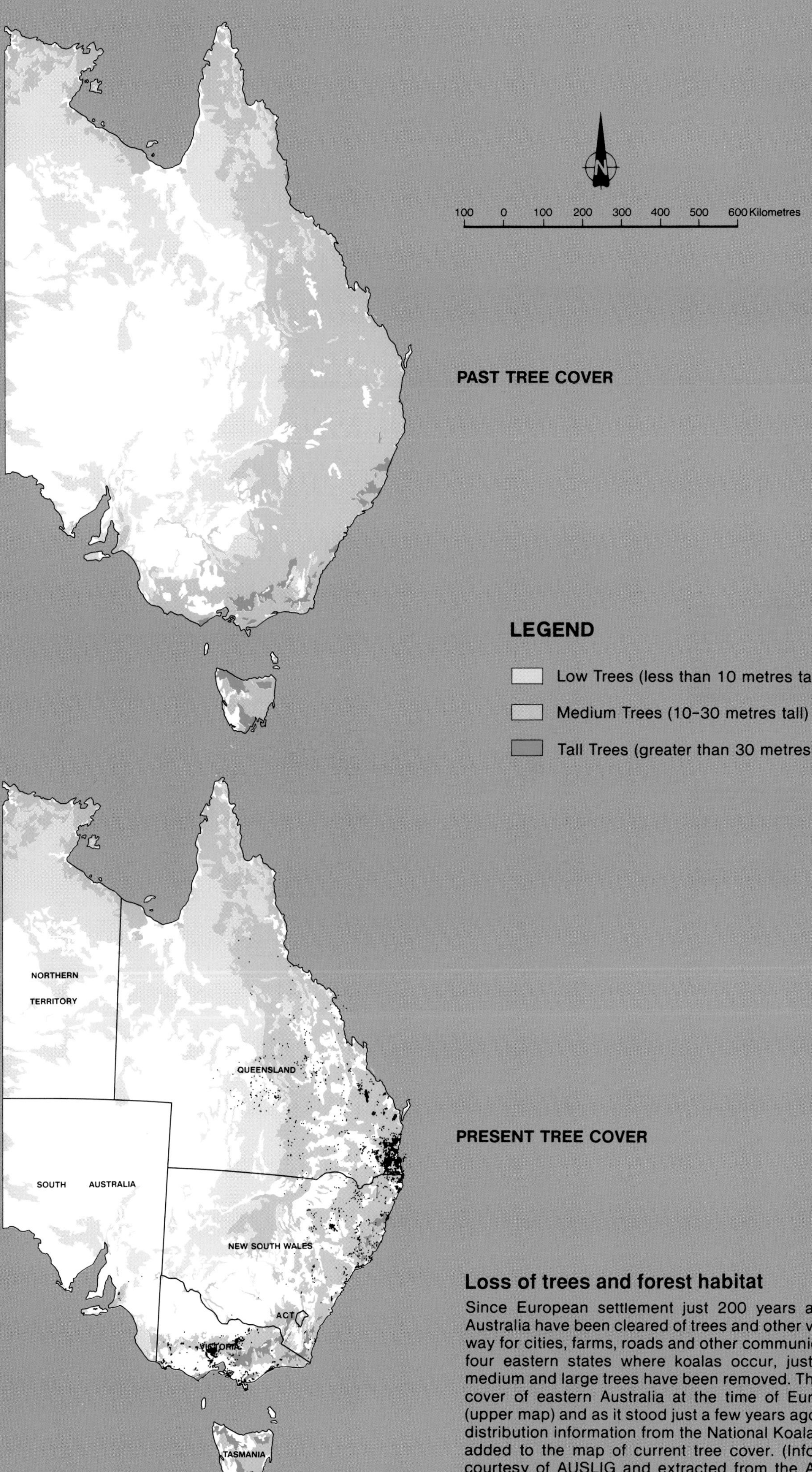

Loss of trees and forest habitat

Since European settlement just 200 years ago vast areas of Australia have been cleared of trees and other vegetation to make way for cities, farms, roads and other communication links. In the four eastern states where koalas occur, just over half of the medium and large trees have been removed. The maps show tree cover of eastern Australia at the time of European settlement (upper map) and as it stood just a few years ago (lower map). The distribution information from the National Koala Survey has been added to the map of current tree cover. (Information provided courtesy of AUSLIG and extracted from the Atlas of Australian Resources, Third Series, Vegetation, in press).

109

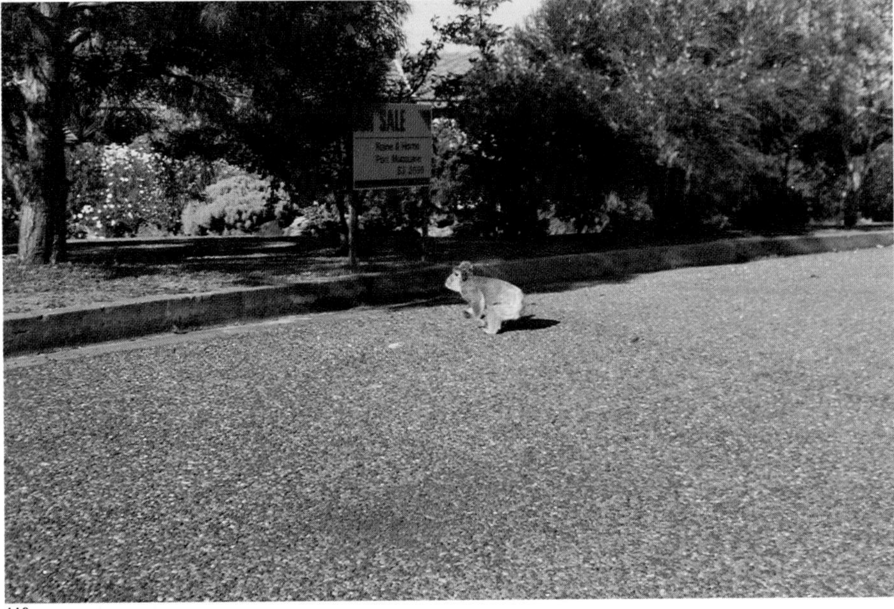

110

109. Koalas have little road-sense.

110. Near urban areas koalas in search of mates or tastier food trees frequently become disoriented.

Why Did the Koala Cross the Road?

Koalas occasionally come down from their leafy refuges and move across the ground in search of mates or more flavoursome food. At these times they may encounter a variety of hazards. Before Europeans settled in Australia the greatest threats to koalas moving from tree to tree were Aborigines and dingoes. These threats have now been replaced by motor vehicles and domestic dogs.

The koala road toll is of greatest concern near the rapidly developing urban centres of coastal Victoria, New South Wales and Queensland. Canfield (1987) found that for the Port Macquarie region of New South Wales, some 30 per cent of dead koalas brought to him for autopsy died from injuries caused by motor vehicles.

The impact of motor vehicles on koalas nationwide is clearly significant although impossible to quantify. On Phillip Island in Victoria, wildlife authorities fight a constant battle against speeding motorists. The Phillip Island koala population has declined alarmingly over the past fifteen years and the high road toll is seen as a major cause.

Backhouse and Crouch (1988) have reported that since 1984, between 40 and 60 per cent of koala deaths each year on Phillip Island are caused by motor vehicles. To reduce this mortality, warning signs have been erected at key road-crossing areas. Near the beginning of this zone a sign notifying motorists of the annual koala road toll has been erected in an attempt to prick the consciences of those exceeding the recommended speed. The effectiveness of this approach remains to be seen.

While it is inappropriate to apply the koala road mortality figures from Phillip Island to the rest of the country, it is clear from other studies that motor vehicles pose a serious threat wherever major roads pass through areas occupied by koalas.

Koalas in the Suburbs

Koalas moving between native forest areas sometimes encounter domestic dogs. Although they do try to defend themselves, sadly koalas are no match for large domestic dogs. The prevalence of koala injuries and deaths resulting from altercations with dogs is growing and, in urban areas adjoining forest habitat, individual or packs of uncontrolled dogs have a serious impact on koala populations.

Largely as a result of harassment by dogs, the sight of a koala sitting atop a telegraph pole or adorning a roadside billboard is becoming more frequent. Disoriented or frightened koalas are increasingly appearing in

111. These days koalas turn up in the strangest of places!

112. It is a sad reflection of modern society that koalas can be the target of shooters.

113. An x-ray of a koala riddled with shotgun pellets.

such unnatural places as schools, cemeteries, racing venues, garbage dumps and construction sites. A recent report tells of shoppers returning to their car to find a koala refugee on the rear seat!

Some of the incidents involving koalas are amusing, but the fact that they are becoming more common is cause for grave concern. The forays of koalas into urban society are fraught with dangers and wildlife authorities have the sometimes difficult task of relocating lost and confused koalas to areas where they will be safe.

Suburban backyards can also pose a threat to koalas. Those koalas seeking refuge in garden sheds, for instance, may encounter pesticides, paints, and a variety of other dangerous substances. Although able to swim several koalas drown in in-ground swimming-pools each year. Modern fences not made from timber posts may prove impassable barriers. The average backyard can be a very unhealthy place for koalas.

Jirrahlinga Wildlife Sanctuary's Mrs Tehree Gordon with one of the victims of the vicious attacks at the Brisbane Ranges National Park.

• Picture: JANINE EASTGATE

Koalas easy hit for cowardly shooters

SOMEWHERE in the 7500-ha Brisbane Ranges National Park near Geelong, a cowardly shooter stalks an easy prey.

Hundreds of bush tracks make it almost impossible for the two park rangers to find the corpses.

The victims are Australia's most lovable ambassadors — koalas.

By SUE HEWITT

At the Jirrahlinga Wildlife Sanctuary yesterday, where two of the victims are recovering, it was obvious that the slow moving koalas presented shooters with little more challenge than a tin can or a street sign.

The sanctuary's Mrs Tehree Gordon has measured the shooter's "handiwork" in the six bullets dug out of one animal and the bullet that broke another's leg.

A third koala had to be destroyed — its hind legs were blown off by a shotgun blast.

The Conservation, Forests and Lands assistant regional manager (operations), Mr Ray Danks, said six koalas had been killed and several injured.

"But there is 7500 hectares of park and what we don't know is how many more carcases there are out there.

"We've got two full-time rangers and they don't expect to have to be looking for this type of thing — they are there to help people enjoy the park.

"When you get something like this (the shootings) going on for several weeks, it's very distressing.

"There are several hundred kilometres of road and tracks and they (the shooters) could be anywhere."

He said the estimated 100 visitors in the park at any time had a better chance of finding the shooters.

"At night you can hear a shot but then you have to chose been 10 different tracks and 10 different directions, and you may not even see any lights."

He was unsure whether more than one shooter was involved. The animals have been shot by both rifle and shotgun.

"It is upsetting to imagine why anyone would shoot a koala — they freeze as soon as you put a light on them at night."

The two shooting victims add to the collection of 15 koalas and 250 other animals and birds being cared for at the park.

The shootings have come when Mrs Gordon is trying to organise a koala awareness campaign to protect them from road deaths, attacks by dogs and other dangers.

112

111

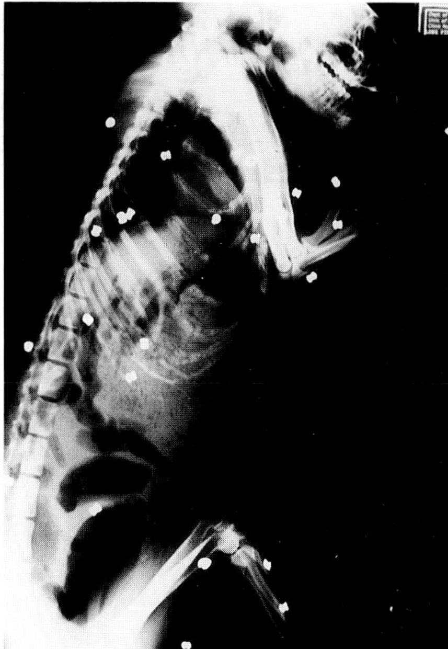

113

Other Threats to Koalas from Society

From earliest times, koala hunting was considered barbaric by some as they were such easy quarry for those with firearms. Despite the protective status which koalas now enjoy, each year there are grisly accounts of koalas maimed or killed by vandals with guns. It is a sad reflection of modern society that defenceless animals can be so cruelly treated. It is impossible to imagine what motivates such senseless action. Shooting koalas could never be considered either a challenge or a sport.

Although the day-to-day threats to koalas from modern society are considerably less in rural areas there are still some dangers for them. Koalas moving across the ground occasionally encounter rabbit traps or baits intended to eradicate feral or pest species. The spraying of insecticides on or near trees occupied by koalas may also contribute some deaths.

114. Koalas spend nearly 80 per cent of their day sleeping or resting.

115. Koalas are generally solitary animals.

116. During the warmer months of the breeding season koalas may be seen on the ground.

A DAY IN THE LIFE OF A KOALA

Koalas are well known for their lethargy spending as much as nineteen hours of every day sleeping or resting. Such prolonged periods of inactivity require little energy and allow koalas to survive on the relatively low-energy diet of eucalypt leaves.

Except for occasional grooming of the fur during the day, most koala activity takes place around dusk or dawn or at intervals during the night. Feeding occupies between one and three hours of a koala's average day and normally occurs in bouts lasting from twenty minutes to two hours. Koalas commonly feed just before sundown and resume feeding a few hours after dark. They may also dine in the hours leading up to dawn.

The uncomplicated lifestyle of koalas is only occasionally interrupted by such exertions as moving from tree to tree, although this movement depends on the density of the forest and the time of year. Where the forest canopy is more continuous, koalas move freely around their home range in search of suitable food without having to descend to the ground. Even in moderately open forests koalas can sometimes avoid ground travel by jumping from branch to branch, something at which they are adept.

Koalas are generally solitary animals but become more mobile in the warmer months of the breeding season. The size of a koala's home range and the extent to which it overlaps with the home range of a neighbour depends on the number of koalas in the area competing for the suitable food trees. On French Island, in Western Port Bay, Victoria, Mitchell (1988) found that the home ranges of koalas were between 1 and 1.5 hectares with considerable overlap of individual areas. In the more widely distributed koala population on Kangaroo Island, South Australia, Eberhard (1972, 1978) reported home ranges up to 2.5 hectares with accordingly less overlap.

On French Island, where koala home ranges are clustered into small groups around patches of large trees, Mitchell (1988) found no evidence of territorial behaviour, even among males with overlapping home ranges. Instead he found that a hierarchy based on dominance operated within the population. For each discrete cluster of koalas there was a dominant male who occupied the largest home range and generally had priority access to the females during the breeding season. The dominant male rushed to evict other amorous males from the trees of 'his' females and even stayed nearby afterwards to ensure the intruder did not return.

Male koalas reach their prime when about five years old and probably do not live much beyond ten. Young males are usually capable of breeding from age two but are rarely given the opportunity by the older, larger males. Once the young males leave the home range of their mother, usually when around two years old, they are generally nomadic for two to three years before establishing their own territory. Young females also depart the home range of their mother when around two years old, but commonly establish a home range close by.

Although it is generally the younger, nomadic male koalas or older, reproductively active ones which are found moving across the ground, the females may also go in search of males during the breeding season or better food supplies. Bellowing males can be heard several hundred metres away in open forest and Lee and Martin (1988) have suggested that in sparse populations this helps receptive females to locate males. They have cited one case where a female koala moved 2 kilometres during the breeding season, ostensibly, to be nearer males.

Despite their apparent lethargy koalas are surprisingly mobile. Records of koalas moving several kilometres are quite common. Robinson (1978) reported that koalas translocated from Kangaroo Island to the Eyre Peninsula, on mainland South Australia, dispersed as far as 29 kilometres from their site of release, in a number of cases across open ground. When walking on the ground, koalas have an awkward gait and bound if forced to hurry. There are a number of anecdotal accounts of koalas swimming, although no detailed description has been published.

114

115

116

NATURAL THREATS TO KOALAS

Old Age

The oldest known koala was a female from San Diego Zoo which lived to be eighteen years old. However, it is unlikely that such an age is rarely if ever achieved by koalas in the wild. Lee and Martin (1988) have reported a female koala from French Island in Victoria which was still breeding when ten years old, and a male from Walkerville which was estimated to be sixteen years old when he died.

These same authors have noted that koalas which escape the other potential hazards in their life may ultimately 'succumb from an inability to masticate sufficient food to meet their nutritional needs'. They base this view on occasionally finding koalas 'with severely worn teeth sitting at the base of a tree seemingly too weak to climb'.

Apart from dying of old age there are several other natural causes of mortality in koalas. Setting aside the sometimes fatal effects of the bacterium *Chlamydia psittaci*, the major natural threats to koalas in Australia today are bushfires and drought.

117

Bushfires

While it seems koalas may survive fires restricted to the leaf litter and shrub layer of the forest, they are not especially mobile and stand little chance of surviving large bushfires, especially those which are propelled at high speed across the forest canopy. In 1898, a massive forest fire in the Gippsland area of Victoria burnt for nearly a month and is believed to have significantly reduced the size of the koala population. More recently, the Ash Wednesday bushfires of 1983 resulted in a decline in the number of koalas at several locations in the Mount Lofty Ranges of South Australia.

Regeneration of fire affected areas is slow and they remain unsuitable for koalas for many years. After five years koalas have only just begun to reoccupy the areas of Horsnell Gully Conservation Park burnt during the Ash Wednesday fires.

118

117. Bushfires, especially when they move at high speed through the forest canopy, are a great threat to koalas.

118. Tree regeneration after a severe bushfire can take several years.

Droughts

The effects of drought are most likely to be felt by koala populations living

119 & 120. In inland areas, koalas live almost exclusively on the trees bordering the permanent watercourses such as Mungallala Creek in arid, south-west Queensland. When the Mungallala Creek dried up in 1979–80 more than 60 per cent of the koala population died.

121. A koala takes shelter during the heatwave, Mungallala Creek, 1979–80.

122. While adult koalas have few natural predators, it is thought that young koalas may occasionally be taken by Powerful Owls and Wedge-tailed Eagles.

119

120

121

in the more arid areas west of the Great Dividing Range. In this region koala distribution is tied closely to the watercourses. An extreme example of the effects of drought on koalas is provided by the dramatic crash of the population living along Mungalalla Creek in south-west Queensland in 1979–80.

Gordon, Brown and Pulsford (1989) have reported that at least 63 per cent of the Mungalalla Creek population died of malnutrition and dehydration in the wake of heat wave and drought conditions. For several seasons preceding the drought, conditions had been good and the koala population had grown to such an extent that less senior members of the population had moved into marginal habitat. The drought and extreme heat resulted in much of the creek drying up and 'caused extensive leaf fall and/or browning of the foliage in food trees'. The more dominant koalas, with home ranges positioned near the permanent water holes, survived.

Such high koala mortality does not always accompany drought conditions, even in the most arid parts of their range. Mungalalla Creek is a rare case and it was precipitated by unusually hot, dry weather following several good years which had allowed the koala population to grow beyond the normal carrying capacity of the habitat.

Natural Predators

Koalas have few natural predators. In 1920, Coverdale observed that dingoes were 'the deadly enemies of the bears, killing numbers of them

122

123. Dingoes are no longer major predators of koalas.

124. In situations where growing koala populations are unable to disperse, such as on islands, they can sometimes literally eat themselves out of house and home.

125. Death from starvation, Quail Island, Victoria, 1943.

when the latter came down to change trees'. Europeans have largely discouraged dingoes from areas supporting koala populations so that nowadays they are no longer considered a significant predator. Young koalas are probably taken by powerful owls (*Ninox strenua*) and wedge-tailed eagles (*Aquila audax*).

Parasites and Diseases

In some areas koalas become heavily loaded with ticks which can cause serious debilitation and also lower their resistance to other conditions. Obendorf (1983) reported mange in two koalas from Victoria and internal parasites, such as tapeworms, have also been found in koalas.

Digestive tract disorders, cancers and pneumonia, resulting from infection by a number of different pathogens, have also been noted in koalas.

Themselves

Where dispersal from an area, such as an island, is not possible, koalas can be their own worst enemy. If population growth is uncontrolled, defoliation of the food trees may result, causing starvation and death.

Defoliation of trees is a potential problem in any area where dispersal avenues are not available to young koalas seeking to establish their own home range away from those of their parents. As koala populations are becoming more isolated due to rural activities and urban expansions, this problem is likely to become more common. Increasingly, mainland populations are being forced into remnant patches of forest with vast expanses of open country or residential housing barring the way of migrating koalas.

Curiously, overcrowding has not become a problem on the several off-shore islands of Queensland which have koalas. The factors responsible for controlling the size of these populations are not known. Some of the island colonies are infected with the bacterium *Chlamydia psittaci* but, so far as is known, this is not believed responsible for their relatively slow growth rate.

123

124

125

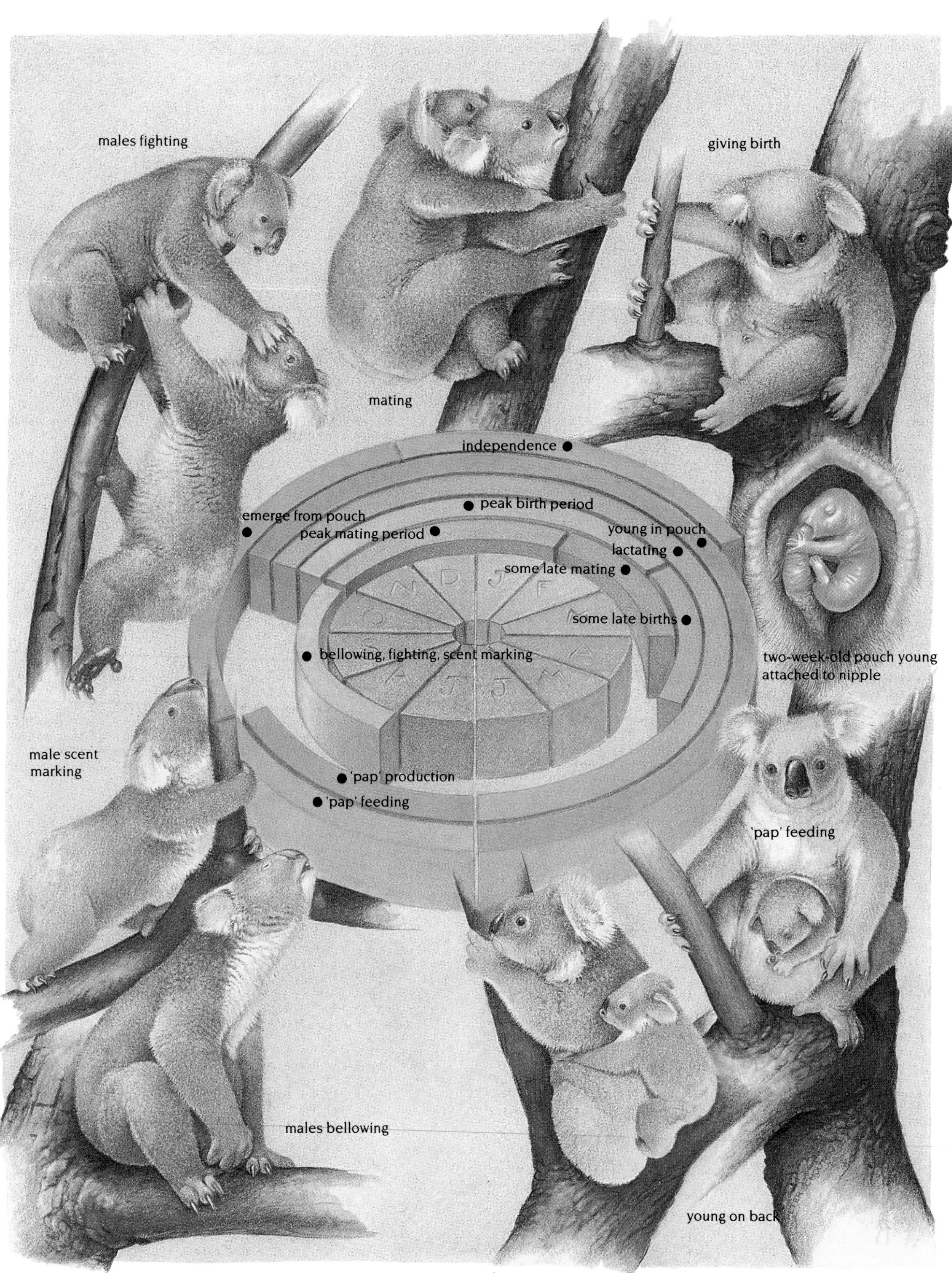
males fighting
mating
giving birth
independence
peak birth period
emerge from pouch
peak mating period
young in pouch
lactating
some late mating
N D J F
some late births
bellowing, fighting, scent marking
two-week-old pouch young attached to nipple
male scent marking
'pap' production
'pap' feeding
'pap' feeding
males bellowing
young on back

126. The breeding cycle of koalas.

127. In koalas, birth takes place thirty-five days after conception.

128. Once in the pouch the young attaches to the nipple where it will stay for thirteen weeks. This young is less than one week old.

129. Even in the newly-born young the forelimbs are well developed and equipped with claws to help them make their way to the pouch. This cub is several weeks old.

REPRODUCTION — KOALA STYLE

With the onset of warmer spring weather the sound of male koalas bellowing in the night becomes more frequent and heralds the start of another breeding season. Although the bellows of the male koalas advertise their whereabouts to females and any competing males, the time of mating is determined by the females. In early spring, most mature females are still providing milk to their offspring of the previous year and until the cub is weaned the next breeding cycle cannot commence.

The time from conception to weaning for koalas is usually around twelve months, with the period of births extending from October through to April depending on the region. In southern Victoria most births occur from December through to February, whereas in the milder climate of Queensland breeding activities begin about a month earlier.

From detailed studies in Victoria, Handasyde (1986), has found that koalas have an oestrous cycle of around thirty-five days. Although koalas are described as seasonally polyoestrous (meaning they have the potential for several oestrous cycles during the breeding season), Handasyde has found that those females whose pouch young perish, for whatever reason, do not breed again until the following year. The fecundity of mature healthy koalas is generally high and it appears most females are successfully fertilised during their first oestrous cycle of each breeding season.

Birth is thirty-five days after conception. Koalas normally have a single young although twins are occasionally born. During birth the mother sits, bent over with hind legs splayed as the tiny newborn makes its way to the pouch and attaches to one of the two teats. Koalas are born without fur, weigh about 500 milligrams and measure less than 2 centimetres. They make their own way from the cloaca to the pouch, pulling their way through the fur using well developed forelimbs equipped with claws.

Koala infants stay attached to the nipple for thirteen weeks at which time the first signs of fur appear. By

127

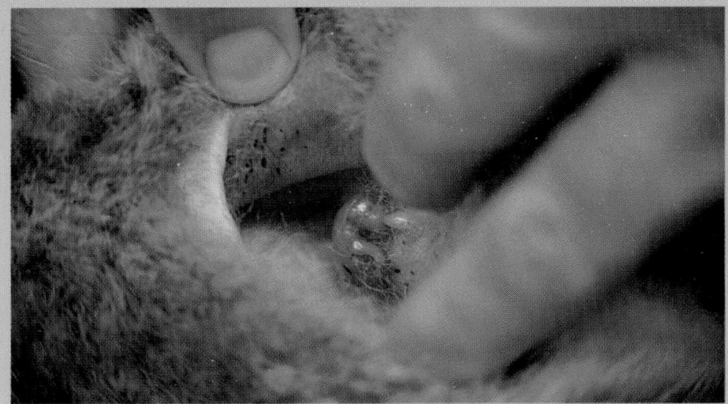

128

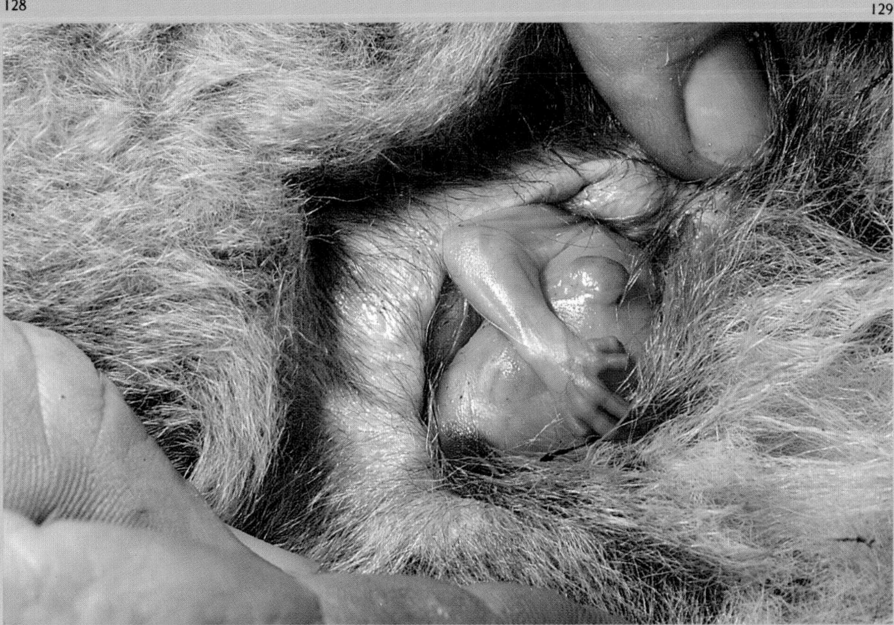

129

twenty-two weeks the young koalas eyes have opened and they take their first cautious glances from the pouch. At this time the mother produces from her anus a dark-coloured fluid called 'pap'. This is thought to come from the mother's caecum, and innoculates the digestive system of the young koala with the microbes which will assist in the breakdown of eucalypt leaves.

At twenty-four weeks infant koalas are fully furred, albeit with darker coloured fur to that which they will sport as adults. Around this time the first teeth appear and they try bits of eucalypt leaves for the first time. At thirty weeks the fur has taken on adult colours and cubs spend lengthy periods clinging to the belly fur of their mothers. Around thirty-six weeks young koalas stop entering the pouch and frequently adopt the characteristic position of riding on their mother's back. By this stage the weaning process has well and truly begun. The frequency of suckling declines as cubs consume greater quantities of leaves.

Once young koalas permanently leave the pouch they slowly begin to move further and further away from their mother. This is a long, slow process. Some eight to ten weeks may elapse before young koalas will venture much beyond a metre away from their mother.

When a cub is weaned, the female koala begins another breeding cycle and the birth of a new suckling young quickly severs any remaining bond with the cub of the previous year. The yearling koalas generally remain in their mother's home range for another year before moving off. The young female koalas establish a home range, sometimes adjacent to or overlapping the home range of their mother or father. In such cases female koalas may eventually be courted by their father. Most young female koalas breed for the first time when about two years old. It seems likely that healthy females continue to produce one young each year until ten to fifteen years old.

In contrast to the young females, young male koalas roam for two to three years after leaving the maternal home range. They don't establish a home range and join the reproductively active koala population until they are physically large enough to compete for females. This is when they are about five years old.

130

131

132

130. When the cubs are about twenty-two weeks old they begin to poke their head out of the pouch. In preparation for digesting the leafy diet, the young koalas eat the microbe-laden 'pap' produced by their mothers.

131. At twenty-four weeks infant koalas are fully 'furred'.

132. The fur changes to adult colours by thirty weeks and the young spends long periods clinging to the belly fur of its mother.

KOALAS AND CHLAMYDIA

Over the past five to ten years much media attention has been given to claims that the bacterium *Chlamydia psittaci* will bring about the extinction of koalas by the end of this century. The accuracy of this claim has now been challenged by a number of researchers who have investigated the *Chlamydia* bacterium and how it affects koalas. These findings are considered in the following section.

The History of Koalas and *Chlamydia*.

It appears that *Chlamydia psittaci* infection in koalas is not a recent happening. Epidemics which decimated koala populations in 1887–89 and 1900–03 are now considered by some to have been caused by *Chlamydia psittaci*. Troughton (1941) reported that the disease was 'some form of opthalmic disease and periostitis of the skull'. In Queensland, during the late 1920s and early 1930s, koalas also suffered the effects of an unknown disease which, according to anecdotal accounts quoted by Gordon and McGreevy (1978), left them 'sitting weak and ill, or lying dead, at the base of trees'.

Early this century, Professor J. P. Hill of the University of Sydney, without being aware of the cause was the first to describe one of the effects of *Chlamydia psittaci* infection in female koalas. He noted 'that the occurrence of cystic ovaries in this species far more frequently than in any other is certainly worthy of record'. The koalas examined by Hill were from south-east Queensland. Following the publication of Hill's observation by O'Donoghue in 1916, Sir Colin Mackenzie noted similar aberrations in the reproductive tracts of koalas from Victoria.

The first published account linking reduced fertility in koalas with reproductive tract disease was by Backhouse and Bolliger in 1961. They found the disorder to be common among koalas around Sydney and that the number of females with young had declined.

Despite continuing interest in the diseases affecting koalas, it was not until 1974 that Cockram and Jackson showed that *Chlamydia psittaci* was responsible for conjunctivitis in koalas. Four years later Dickens suggested that the same organism may be responsible for the cystitis observed in koalas. In 1981 Martin reported reduced fertility among the female koalas at Phillip Island and Walkerville in Victoria and, shortly afterwards, McColl and associates (1984) implicated *Chlamydia* as the organism responsible. This was confirmed by Brown and Grice in 1984.

Besides infecting the reproductive tract of koalas and causing conjunctivitis, *Chlamydia psittaci* is also responsible for urinary tract disorders and has been implicated as a possible cause for pneumonia/rhinitis. All of these conditions were noted by Noel Burnett, proprietor of

WHAT IS CHLAMYDIA PSITTACI?

Chlamydia psittaci (Clam-mid-ee-a sit-a-ki) is a tiny bacterium which spends most of its life within the cells of the host organism. During the 1930s, *Chlamydia psittaci* was responsible for a minor epidemic of pneumonia among elderly people in Europe and the United States of America. The disease was referred to as 'parrot fever' because the primary hosts for the bacterium were cage birds, such as parrots and budgerigars, commonly kept as pets by the aged.

Subsequent research found that *Chlamydia psittaci* was widespread and common in birds and a number of mammals. Brown and Carrick (1985) have reported that the bacterium causes a variety of conditions in domestic pets and livestock, including conjunctivitis in cats, arthritis in dogs and lambs and abortion in sheep and cows.

Chlamydia trachomatis, a similar organism, occurs exclusively in humans where it is responsible for the eye disease, trachoma. This same bacterium also infects the reproductive tract of humans, sometimes resulting in pelvic inflammatory disease and subsequent infertility in women.

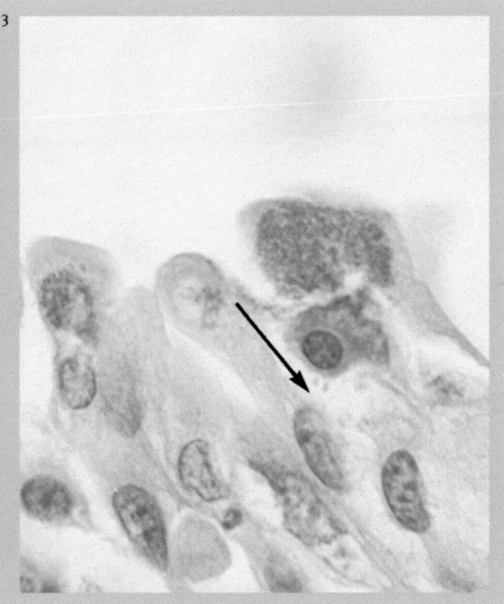

133. This 400 per cent magnification of the cells in the bladder of a koala shows the inclusion bodies (arrowed) of the bacterium, *Chlamydia psittaci*. Inclusion bodies are groups of *Chlamydia* multiplying within the cells.

134 & 135. Koalas infected with *Chlamydia* may be affected in a number of ways. These koalas have *Chlamydia* infection in the form of conjunctivitis — commonly called 'pink eye'.

136. *Chlamydia* can also infect the urinary tract of koalas and results in the soiling of the hindquarters with urine — known colloquially as 'wet bottom' or 'dirty tail'.

137. Frank Carrick of Queensland University supervises the treatment of a koala suffering from conjunctivitis.

the Queensland Koala Park Sanctuary, and reported by Ambrose Pratt in *The Call of the Koala* published in 1937. Burnett found that 50 per cent of the koalas brought to his sanctuary were suffering from 'the condition of cystic reproductive organs' and he suggested that disease would 'eventually wipe out the koala, perhaps within a few years'.

While similar views were expressed by others, it wasn't until Brown and his co-workers published the results of an Australia-wide survey of koala populations in 1984 that concern for the future of the species attracted widespread attention and publicity. During this survey, 237 female koalas were screened using a radiographic technique, and 101 (43 per cent) showed signs of cystic structures associated with the ovaries. Seven of the eight sites where koalas were tested yielded positive results and given the spread of the sites across the range of koalas, Brown and associates concluded that the condition was widespread and reason for grave concern.

The Effects of Chlamydiosis on Koalas

While it appears *Chlamydia psittaci* may manifest itself in koalas in four different ways, the two most obvious are infection of the urinary tract and eyes.

Conjunctivitis, commonly referred to as 'pink eye', may develop in one or both eyes and is characterised by red and swollen eyelids and sometimes there is an eye discharge. Mild infections have been observed to regress of their own accord, but in some severe cases the cornea becomes clouded and vision is impaired.

Urinary tract infection in koalas result's in soiling of the hindquarters by urine and is known colloquially as 'wet bottom' or 'dirty tail'. It is not known whether 'dirty tail' can lead to infection of the reproductive tract or vice versa, although it seems likely given the closeness of the respective openings.

Infection of the respiratory system appears to be the least common form of *Chlamydia* infection in koalas.

134

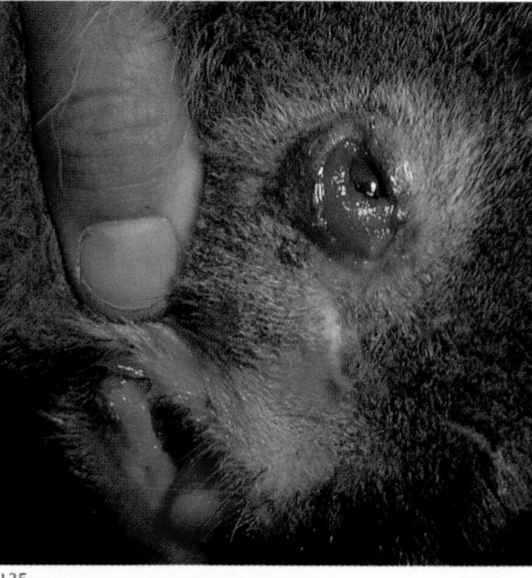

135

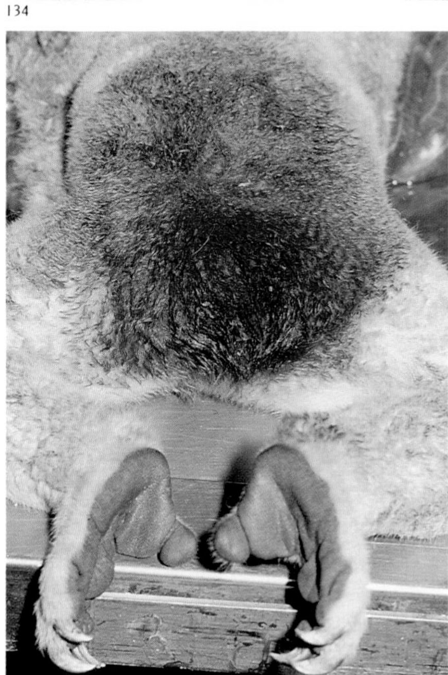

136

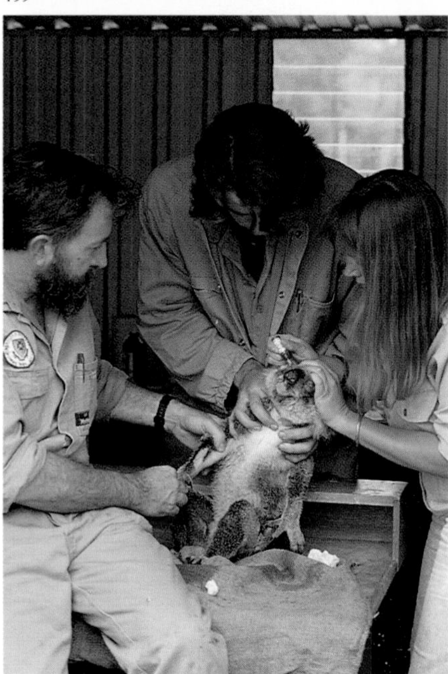

137

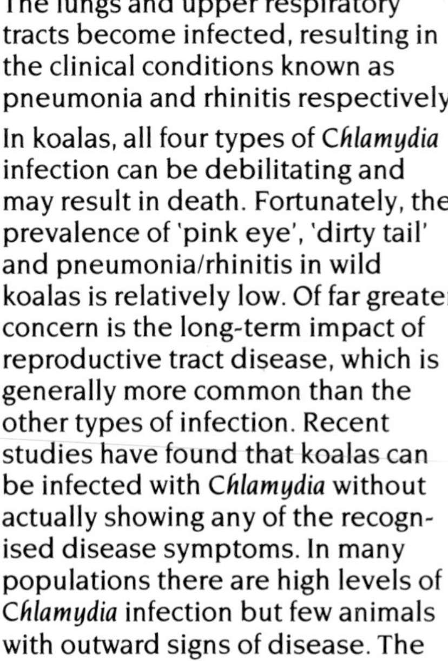

The lungs and upper respiratory tracts become infected, resulting in the clinical conditions known as pneumonia and rhinitis respectively.

In koalas, all four types of *Chlamydia* infection can be debilitating and may result in death. Fortunately, the prevalence of 'pink eye', 'dirty tail' and pneumonia/rhinitis in wild koalas is relatively low. Of far greater concern is the long-term impact of reproductive tract disease, which is generally more common than the other types of infection. Recent studies have found that koalas can be infected with *Chlamydia* without actually showing any of the recognised disease symptoms. In many populations there are high levels of *Chlamydia* infection but few animals with outward signs of disease. The main impact is reduced fertility but often the fertility is adequate to allow the population to continue to expand at a reduced rate.

Treatment for Koalas with *Chlamydia*

There is at present no means by which the bacterium *Chlamydia psittaci* can be eradicated from wild populations of koalas. Unfortunately, antibacterial drugs kill the micro-organisms which koalas rely upon to breakdown eucalypt fibre in the caecum and are often fatal. Captive koalas can be successfully treated with the commonly used antibacterial drug, tetracycline, but depend on intensive care and special feeding to survive.

138

138. Koalas in captivity can be successfully treated against *Chlamydia* using standard antibacterial drugs providing intensive care is given.

139

139. There is strong evidence that *Chlamydia*, which can also infect the reproductive tract, can be transmitted from koala to koala during the breeding season.

Even if drugs or a vaccine specific for *Chlamydia* were available it is unlikely they could be administered to every koala in an infected population in order to prevent reinfection.

How is *Chlamydia* Passed from Koala to Koala?

In humans, *Chlamydia trachomatis* is transmitted by direct contact, as in sexual intercourse or by contact with contaminated surfaces. The unborn child can even be infected when passing down the birth canal. In some birds and mammals, Storz (1971) has suggested ticks and lice may transmit *Chlamydia* or that inhaling infectious particles may bring about pneumonia. Are these methods of transmission feasible for koalas?

Being basically solitary animals the opportunities for *Chlamydia* to be transmitted from one koala to another by direct contact are somewhat limited. During the breeding season, males and females come together briefly to mate, but apart from this, contact between adult koalas is generally avoided. In captive koalas at least, the first attempt at insemination has a high success rate and females apparently do not engage in repeated matings, despite the attempts to do so by amorous males. The only other contact between koalas, apart from fights between males, is that of a mother with her offspring. Infants being weaned eat 'pap' from their mother's cloaca; it is assumed to inoculate their own caecum with the microbes needed to digest eucalypt leaves. Young koalas remain in close contact with their mother for six months after leaving the pouch and do not normally leave her home range until they are about two years old. Clearly there is opportunity for young koalas to become infected from their mothers.

In a survey of koalas around Lismore and Port Macquarie, in New South Wales, Cockram and Jackson (1981) found that 5 per cent of those with conjunctivitis were less than one year old and 18 per cent were in their second year of life. A recent study in the Redlands Shire of south-east Queensland, by Weigler and associates, has found a similar rate of *Chlamydia* infection in immature koalas as in adults.

In the Brisbane Ranges of Victoria, Lee, Martin and Handasyde (1988) reported that thirteen of the sixteen sub-adult koalas tested were *Chlamydia*-positive. *Chlamydia* antibodies were detected in the blood of some of these koalas when they were only eighteen months old suggesting that infection had taken place three to four months earlier. Since this is far too young for any sexual activity and long after any intimate contact with their mothers, these researchers suggest that the immature koalas tested may have possessed antibodies produced in response to the conjunctivitis form of *Chlamydia* infection.

The key to unravelling this mystery may lie in finding of Girjes and associates (1988) that the conjunctivitis form of *Chlamydia psittaci* is a different strain to that infecting the urinary and reproductive tracts. Lee, Martin and Handasyde (1988) speculate that some arthropod vector, such as the bush fly, may be responsible for transmitting the ocular form of *Chlamydia*, notably to sexually immature koalas. In areas where ticks are present, they also may play a role in the transmission of *Chlamydia*.

Is *Chlamydia* a Sexually Transmitted Disease?

There are now strong indications that the strain of *Chlamydia* which infects the reproductive tract of koalas is sexually transmitted. In 1986, Handasyde, working in Victoria, provided circumstantial evidence that *Chlamydia* was passed from koala to koala during sexual intercourse. Koalas without *Chlamydia* which were introduced into an infected population remained free of infection up until the onset of the next breeding season.

More recently Handasyde, in collaboration with Lee and Martin, has provided stronger evidence for the sexual transmission of *Chlamydia* in koalas. These researchers placed *Chlamydia*-free females and *Chlamydia*-infected males on Chinaman Island, a small koala-free island in Western Port Bay. Within twenty-one days the sexually active females became infected with *Chlamydia*. The following year two uninfected male koalas were released into the now infected Chinaman Island population and developed reproductive tract disease during the breeding season.

For further evidence, these same researchers introduced infected and uninfected female koalas onto another island, Churchill Island. In the absence of males there was no transmission of the disease from the infected to the uninfected females.

KOALAS COMMUNICATING

Because koalas tend to live separately, they have a relatively simple social system. Nevertheless, they do communicate with one another using scent marking and vocal calling.

The scent glands in the centre of the chest of male koalas produce a pungent, orange-coloured secretion which is rubbed on the base of trees and along branches. No doubt this provides a social signal to other koalas. The degree to which scent marking indicates to other koalas the status of the animal responsible for it is not yet known.

Another form of scent marking is used by both male and female koalas. Urine is dribbled on the trunk of trees and sprinkled on the ground nearby. Again, the precise significance of this behaviour is unknown, although it is likely that other koalas are made aware of the reproductive status of the koala in the tree above by the odours coming from the urine.

Koalas also make a variety of calls. Some calls are important in maintaining the social spacing within a koala population. Both male and female koalas bellow, those of males being generally louder and more frequent.

The males bellow consists of a tremulous inhalation of air followed by a louder but shorter exhalation. Smith (1980) has referred to these as the snore and the belch phases. The bellow can last from ten to thirty seconds and depending on the openness of the forest may carry for several hundred metres.

Fights between males are not common but they are violent when they occur. Older males bear the scars of past battles, which are a frenzy of biting and scratching concluding with a scream from the vanquished and a triumphant bellow from the victor.

From his extensive observations of the behaviour of koalas in captivity, Smith (1987) has concluded that 'females are more aggressive but not as violent' as males. Female koalas use a suite of calls to communicate their attitude to an impending confrontation. Rather than fighting, a snarl is used in an attempt to intimidate potential opponents. If aggression is overtaken by fear, female koalas issue an almost plaintive wail and will scream if seriously threatened.

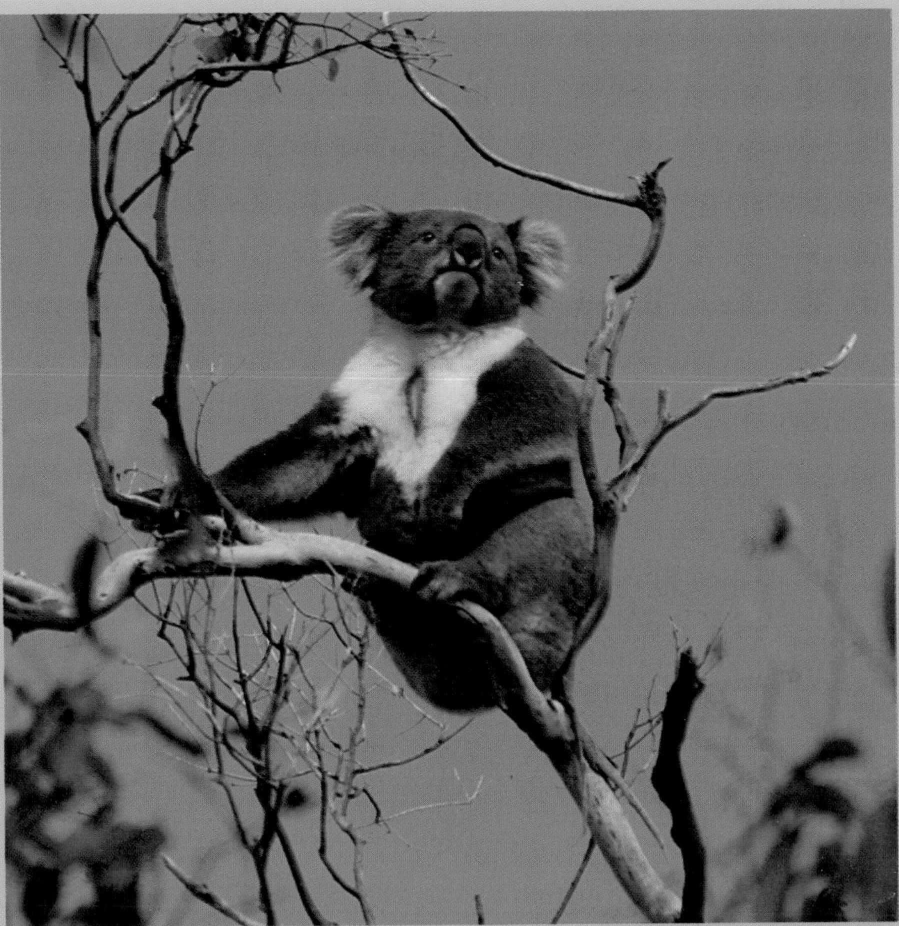

140

Mating behaviour in koalas is also associated with considerable vocalisation from both males and females. When the female is in oestrous (on heat) several males may attempt courtship. The female greets the approach of a possible suitor with snarls and bellows. If the male is deemed unsuitable the female may scream as he attempts to mate. She escapes by biting the male, who either remains in the tree to try again or reluctantly backs down the tree bellowing as he goes.

Should the screams from unwilling females be heard by the dominant male from the area, he soon arrives and expells the intruder. Sometimes he remains nearby, on guard.

When mating takes place it generally lasts for no more than two minutes. The male mounts the female in a vertical position and bites onto the back of her neck. Through the female stretching her neck back and lifting her rear the male gains intromission and a vigorous series of pelvic thrusts follow. Mitchell (1988) has reported that following this, abdominal contractions occur in both the male and female before she separates from the male, sometimes uttering a low squawk. The male bellows as he moves back down the tree.

The simplicity of the koala social system is reflected in the interactions between mother and young. Female koalas are mostly indifferent to their offspring; they only occasionally lick them and show only mild interest in the squeaks of alarm their offspring sometimes make. In captive colonies, cubs are frequently swapped between parents but this is unlikely to occur in the wild because of the solitary nature of koalas.

141

142

143

140. A male koala displays the pungent, orange-coloured secretion from the glands on his chest.

141. A male koala marking a branch with the secretions from his chest glands.

142. Both male and female koalas have calls. These include bellows, snarls and screams.

143. Fights between male koalas are not common but are a frenzy of biting and scratching.

The Chinaman Island and Churchill Island studies have provided good evidence that *Chlamydia psittaci* is sexually transmitted between koalas. What is still not clear is whether or not there are other ways the bacterium is transmitted and, if so, how significant they are.

In their Chinaman Island study, Lee, Martin and Handasyde (1988) noted that two of the three females which became infected with *Chlamydia* as a result of mating with infected males, subsequently lost the young produced from this union. The cause of these deaths was unknown but the researchers have suggested that the unborn young may have become infected while moving down the birth canal of their mothers, just as sometimes occurs in humans infected with *Chlamydia trachomatis*. However, the very low prevalence of *Chlamydia* infection in juvenile koalas suggests that this may not be a common method of infection, if at all.

How Common is *Chlamydia* in Wild Koalas ?

One aim of the National Koala Survey was to collect information on the prevalence and distribution of *Chlamydia* in wild koala populations. The results of this part of the survey must be interpreted with extreme caution as koalas perched high in trees are not easily observed and some instances of disease may have been overlooked. Equally, not all infected koalas show outward signs of disease and the early stages of 'pink eye' and 'dirty tail' in koalas are not always readily recognised, even by experienced observers. Despite these limitations, the results of the national survey, when considered with other research information, do allow for an up-to-date appraisal of the current prevalence and distribution of *Chlamydia psittaci* in wild koalas.

Of the 203 koalas reported during the national survey in South Australia, no external signs of disease were noted. During the survey, blood tests were conducted on sixty koalas from Kangaroo Island. None recorded *Chlamydia*-positive results.

In Victoria, a total of 1702 koalas, both live and dead, were sighted in the National Koala Survey. Of these, twenty-seven (1.6 per cent) were reported to have 'pink eye' or 'dirty tail'. This proportion is consistent with surveys of Victorian koala populations by Lee, Martin and Handasyde (1988) who have found that in populations infected with *Chlamydia*, 'pink eye' and 'dirty tail' sufferers represent less than 3 per cent and 4 per cent respectively.

In New South Wales, participants in the national survey were asked if they had seen any sick, injured or dead koalas in the previous year. Reed, Lunney and Walker (1988) record that 159 (22.8 per cent) of the 697 replies noted 'unhealthy' koalas. It is important to note that the 'unhealthy' category includes all injured and dead koalas as well as those suffering from *Chlamydia* infections. From their analysis these authors conclude that 'unhealthy' koalas were found throughout only 8.2 per cent of their range in New South Wales and that the 'much publicised disease issue is not the primary determinant of the distribution of the koala' in that State.

There have been few other studies on the prevalence of *Chlamydia*-induced diseases in New South Wales koalas but in 1981, Cockram and Jackson surveyed koala populations in the Lismore and Port Macquarie areas. They were specifically looking for instances of conjunctivitis. Of eighty-five koalas observed near Lismore, ten showed signs of 'pink eye' and around Port Macquarie none of the forty-nine koalas examined were affected in this way.

In Queensland, during the national survey, some sixty-nine reports out of 1 122 (6.1 per cent) recorded the presence of sick koalas. As in Victoria, the National Koala Survey in Queensland recorded slightly fewer cases of chlamydiosis than more intensive localised studies undertaken in the past few years. During a survey of koalas in south-east Queensland, Weigler and co-workers (1988) examined sixty-five koalas of which six (9.2 per cent) had either 'dirty tail' or 'pink eye'.

In studies of koala populations near Oakey and Springsure in central southern Queensland, Gordon, McGreevy and Lawrie (1989) have monitored the prevalence of 'dirty tail' and 'pink eye' over several years. Conjunctivitis was less prevalent than 'dirty tail' at both sites, generally affecting less than 4 per cent of the population. In most years conjunctivitis was not observed in either population. The prevalence of 'dirty tail' syndrome varied from year to year in both populations, ranging from zero to levels as high as 12 per cent.

What Does it All Mean?

The findings of the National Koala Survey and other studies, indicate that *Chlamydia* infection occurs in koalas across most of their range. In infected populations the number of koalas showing the disease symptoms generally represents less than 5 per cent of those present. If uninfected populations are included then the rate of overt *Chlamydia* infection in the entire koala population is considerably less than this figure and probably approaches 2 per cent.

On face value the low prevalence of 'pink eye' and 'dirty tail' is encouraging. However, the keys words here are 'showing the disease symptoms' for, as stated earlier, researchers have now demonstrated that some koalas carry *Chlamydia* without displaying any outward signs or ill effects.

In their Redlands Shire study, Weigler and associates (1988) found only 9 per cent of koalas showed the commonly recognised disease symptoms even though 71 per cent of animals tested were *Chlamydia* positive. Similarly, in the six Victorian *Chlamydia*-infected populations surveyed by Lee and co-workers (1988), less than 5 per cent of koalas were overtly diseased although infection rates as high as 86 per cent were recorded.

While the rate of *Chlamydia* infection in koalas is apparently greater than outward signs of the disease indicate, it is heartening that in most infected populations there has not been a total shut-down in breeding. Lee, Martin and Handasyde (1988) have provided details of koala populations in Victoria which continue to grow despite high rates of *Chlamydia* infection. In four such populations they recorded fertility rates ranging from 38 per cent to 62 per cent even though *Chlamydia* infection levels were between 56 per cent and 86 per cent. Furthermore, it is now apparent that female koalas in Queensland are also not immediately stricken with infertility as a result of infection with *Chlamydia psittaci*.

Under ideal conditions koala populations can multiply rapidly. Even though a somewhat artificial situation, the rapid population growth of koalas on Quail Island in Western Port Bay, Victoria, vividly emphasises this point. In this idyllic environment the population, free of *Chlamydia*, grew so rapidly that massive tree defoliation occurred. The consequent death of koalas attracted public attention. Martin, Handasyde and Lee (1987) have calculated that in just over a decade the population had grown at a annual rate of 17.5 per cent, on average doubling their population size every 4.3 years.

Experience at Raymond Island in the Gippsland Lakes of Victoria emphasises that koala populations can continue to grow even though infected with *Chlamydia psittaci*. In 1953, forty-two koalas from the *Chlamydia*-infected Phillip Island population were relocated onto Raymond Island. By 1980 the population had grown to 196 representing a annual growth rate of 6 per cent and a population doubling time of 12.2 years.

Long Term Concerns

Despite the evidence that some koala populations infected with *Chlamydia* continue to breed and grow, it is wrong to assume this occurs universally or will continue indefinitely.

As researchers have looked more closely at koalas across the country it has become apparent that the effects of *Chlamydia psittaci* are more pronounced in some populations than in others. Koalas not previously

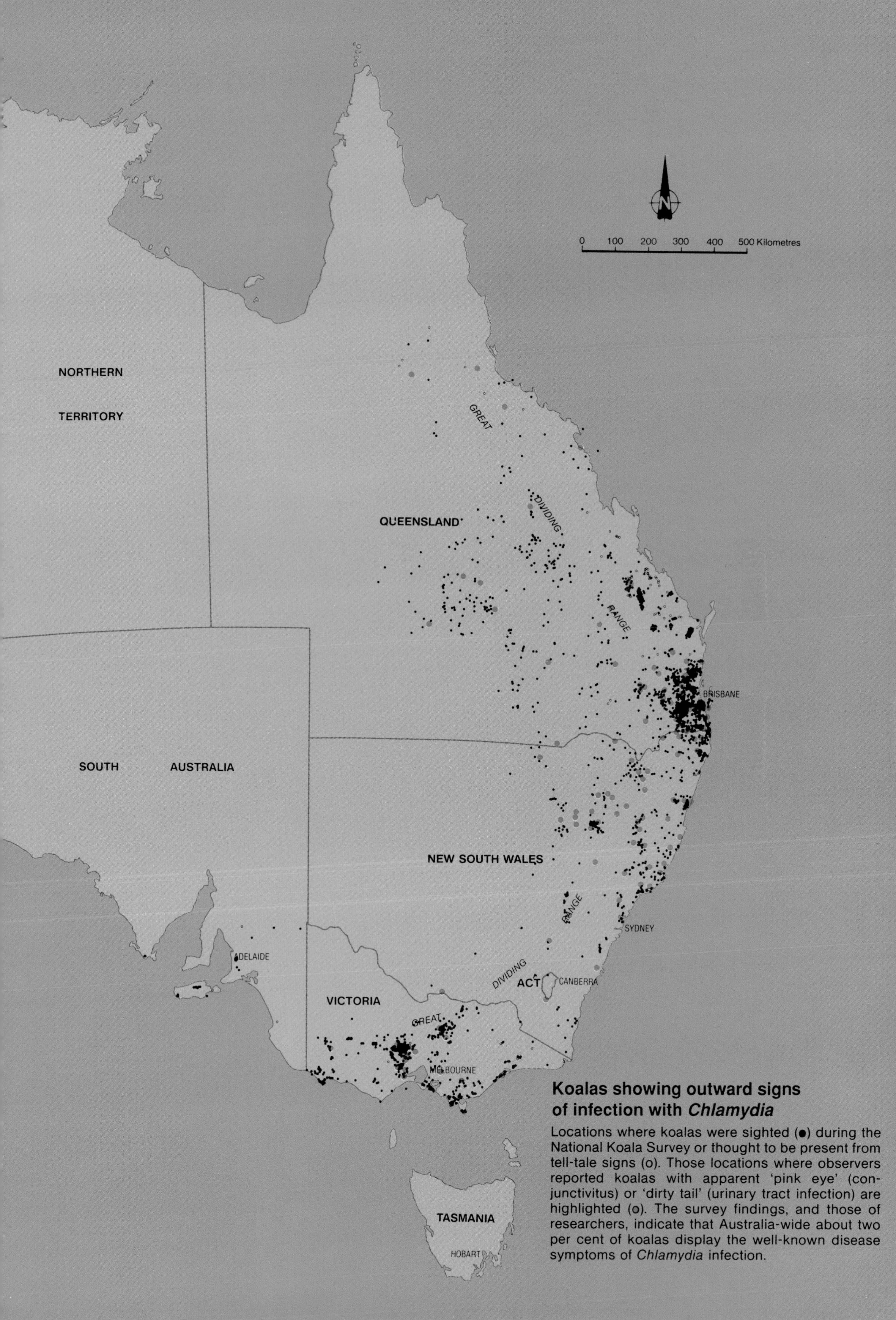

Koalas showing outward signs of infection with *Chlamydia*

Locations where koalas were sighted (●) during the National Koala Survey or thought to be present from tell-tale signs (o). Those locations where observers reported koalas with apparent 'pink eye' (conjunctivitus) or 'dirty tail' (urinary tract infection) are highlighted (o). The survey findings, and those of researchers, indicate that Australia-wide about two per cent of koalas display the well-known disease symptoms of *Chlamydia* infection.

144. In southern Australia koalas are routinely relocated because of overcrowding or direct threat from land development. Great care must be taken not to introduce koalas infected with *Chlamydia* into populations not previously exposed to the bacterium.

145. The koala population on Victoria's Phillip Island has been in serious decline over the past fifteen years due to the combined effects of urban encroachment, motor vehicle related mortalities and disease.

144

NEVER MIND. EVERY SPECIES
FEELS ENDANGERED AT
ONE TIME OR ANOTHER

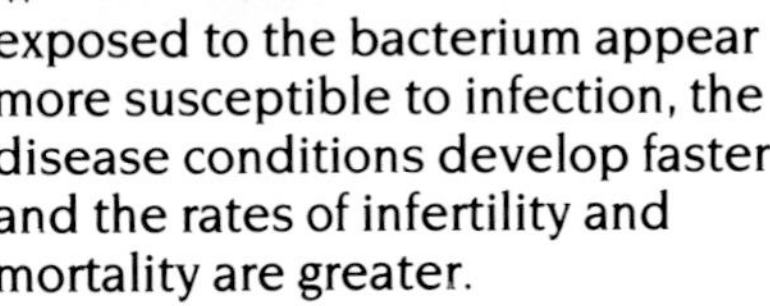

exposed to the bacterium appear more susceptible to infection, the disease conditions develop faster and the rates of infertility and mortality are greater.

In the Grampians National Park, Victoria, koalas were re-introduced in 1957 from the disease-free population on French Island. They flourished until about 1983 when it is thought *Chlamydia* was introduced into this population, possibly through the unauthorised release of an infected koala. Today the Grampians population has an 84 per cent *Chlamydia* infection rate but unlike other infected populations in Victoria, breeding has come to a virtual halt and fertility is considered to be zero. Lee, Martin and Handasyde (1988) suggest that this case illustrates the more severe effects *Chlamydia* has on koalas without previous exposure to the bacterium and comment that '*Chlamydia* does pose a threat to susceptible populations'.

Another Victorian koala population infected with *Chlamydia* which is known to be in decline is that on Phillip Island. Unlike the Grampians koalas, the decline of this population cannot be attributed to the recent introduction of *Chlamydia*, as the bacterium has been present among Phillip Island koalas since at least the 1950s. The prevalence of *Chlamydia* infection in the Phillip Island koalas is currently 83 per cent but the fertility rate is a lowly 17 per cent and the population is in decline. This contrasts with the Brisbane Ranges population, near Melbourne, where the level of *Chlamydia* infection is similar to that on Phillip Island at 78 per cent yet fertility is appreciably higher at 52 per cent and the population continues to grow.

An important difference between the Phillip Island and Brisbane Ranges populations appears to be their exposure to expanding urban centres. The declining Phillip Island population is part of an increasingly urbanised environment in which they are constantly at risk from motor vehicles and domestic dogs. Some researchers contend that under such conditions of harassment, the likelihood of *Chlamydia* disease symptoms appearing in the population is greater. The conditions such as 'pink eye' and 'dirty tail' are now recognised as indicators of an individual or koala population which is under stress. While researchers are yet to define in precise detail what comprises stress in koalas, urban encroachment into their habitat would appear to be responsible on Phillip Island.

Koala Stress Syndrome

From his experiences in treating sick and injured koalas in Victoria, Obendorf, in 1983 described what he called 'Koala Stress Syndrome'. Among the koalas he examined, this was 'characterised by lassitude, depression, anorexia and coma' and was identified in animals brought in from the wild as well as those which required lengthy treatment and hospitalisation. Obendorf found that the sufferers of this condition were mostly mature male koalas. 'They were found wandering aimlessly or prostrate and comatose, with no evidence of trauma or overt illness. Some affected animals were found at the base of trees reluctant to climb or eat'.

It is now believed by some researchers that Koala Stress Syndrome results from hormonal imbalance and breakdown of the immune system, brought about by stress. The relationship between the syndrome and chlamydiosis is still under investigation but it seems to

145

146. Koalas on Phillip Island must live with an increasingly urbanised environment which means 'running the gauntlet' in order to find mates or other habitat areas.

147. The past five years has seen a flurry of research into koalas.

predispose koalas to infection with *Chlamydia psittaci*, or to developing the 'pink eye' and 'dirty tail' conditions if the bacterium is already present in their body.

Overcrowding is another factor apparently playing a significant role in the prevalence of chlamydiosis in koala populations. From studies of koala communities in Queensland, beginning in 1971, Gordon, McGreevy and Lawrie (1989) have concluded there is an association between the prevalence of 'dirty tail' and population density which can sometimes lead to the 'temporary local rarity' of koalas.

In the semi-arid areas of inland Queensland, koalas are present in small patches of favourable habitat, usually along rivers and streams. Rainfall patterns primarily dictate the density of koalas within these patches. During good seasons the population spreads out as drier, marginal habitat becomes suitable, at least for the younger sub-dominant koalas. In drier times the population contracts back into the prime habitat and the routes for dispersing young or misplaced koalas are no longer available. In response to the resultant increase in population density, the prevalence of *Chlamydia* disease syndromes can be expected to increase.

Gordon and co-workers suggest that the association between disease and high population density may help to explain the epidemics which ravaged koala populations in the late 1800s and early 1900s. Anecdotal reports indicate that these epidemics, which many attribute to *Chlamydia psittaci*, were associated with the very high local abundance of koalas.

146

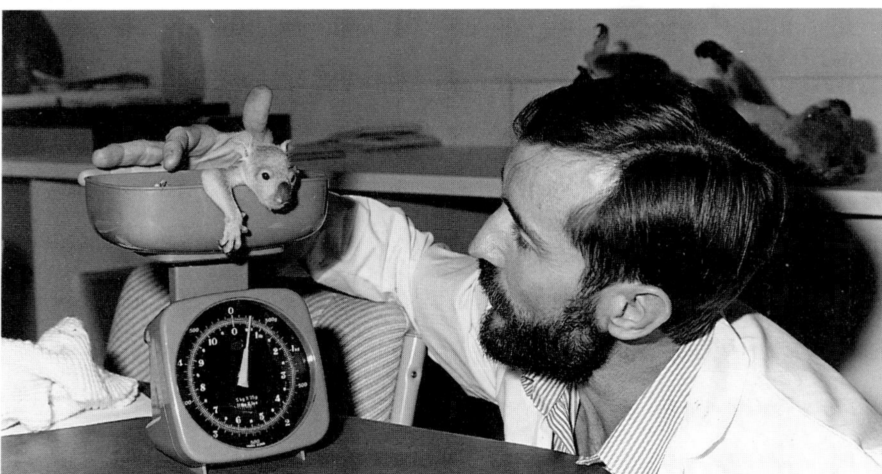

147

Putting the Threat from *Chlamydia* in Perspective

The last five years have seen a flurry of research activity directed towards improving our understanding of the bacterium *Chlamydia psittaci* and how it affects koalas. The findings provide both good and bad news for the future of the species.

On one side of the ledger, it is encouraging that a number of disease-free populations exist and that others continue to breed and grow despite harbouring the bacterium. Weigler and co-workers suggest that this 'indicates a host-parasite relationship that is long-established and generally stable', typical of 'the common type of *Chlamydia* infection in other animal species'.

However, this does not mean that there is room for complacency. There are strong indications that this 'generally stable' relationship between *Chlamydia psittaci* and koalas breaks down with overcrowding, urbanisation and habitat loss. Given the likelihood of continuing urban expansion and habitat fragmentation imposing these pressures on more and more koala populations, appropriate management strategies must be developed as soon as possible.

The vulnerability of koalas not previously exposed to *Chlamydia* is a major concern. While the program of relocating koalas from the islands of Western Port Bay, Victoria, has served to re-establish the species throughout southern Australia, the distribution of *Chlamydia*-infected populations in Victoria, can also be largely attributed to this program. Koala populations must be managed with an awareness of the possible impact of introducing infected koalas into *Chlamydia*-free communities. If they are not there is clearly potential for local koala populations to be significantly depleted by disease.

ARE KOALAS AN ENDANGERED SPECIES?

In 1981, Strahan and Martin presented a paper to a 'Species at Risk' conference in Canberra, entitled 'The Koala: Little Fact, Much Emotion'. This title accurately reflected the situation at the time but marked the beginning of an era of intense research and interest in koalas. Sparked by public concern for the future of the species, research was undertaken into many aspects of koala biology, particularly the effects of the bacterium *Chlamydia psittaci*.

Several years later our understanding of koalas and their needs has improved greatly, but public concern for their future has waned little, if at all. The question still being asked is 'are koalas endangered?'

The International Union for Conservation of Nature and Natural Resources (IUCN) is an authoritative voice on matters of species status. This body describes endangered species as those 'in danger of extinction and whose survival is unlikely if the causal factors continue operating'. Put another way, this includes species 'whose numbers have been reduced to a critical level or whose habitats have been so dramatically reduced that they are deemed to be in immediate danger of extinction'.

From the results of the National Koala Survey and other information, it cannot be said that koalas are in 'imminent danger of extinction'. It is impossible to reconcile their abundance in many areas and widespread distribution with the IUCN definition of endangered. In South Australia and Victoria, island populations continue to grow, management arrangements are in place and surplus koalas are used to seed new mainland populations. In New South Wales and Queensland, despite contractions of their range and population thinning in some areas, koalas remain widespread and in places, abundant.

Claims that the bacterium *Chlamydia psittaci* could be the causal factor responsible for the extinction of koalas are not sustainable on current evidence. The disease syndromes, 'pink eye' and 'dirty tail', are apparent in less than 2 per cent of koalas and most *Chlamydia*-infected populations are able to maintain a slight positive growth rate despite the infertility and mortality associated with the bacterium. Researchers now agree that while *Chlamydia* has the potential to bring about local population declines, this usually occurs in concert with other factors, and does not pose a threat to the continued existence of the species.

But . . .

As reassuring as it is to conclude that on present day evidence the future of koalas on a national scale is secure, it should not be overlooked that some local populations are immediately threatened and likely to be exterminated within the next few years. The questions to be asked are: are these the first signs of a growing problem or just unfortunate, isolated incidents? and, do the problems which some local populations battle against today have the potential to one day pose a major threat to the species? The declining koala population on Phillip Island in

148 & 149. Recent research has allowed for an informed appraisal of the current conservation status of koalas.

150. Although on a national scale koalas are not endangered, there are a number of local populations immediately threatened by expanding urban developments and associated factors.

148

149

151. In areas where the shrubs and other understorey plants are removed by livestock or as a fire-preventative measure, non-eucalypt trees may become established.

152. Roadside trees, which may form part of precious habitat remnants, are also vulnerable to 'dieback' caused by the root rot fungus, *Phytophthora cinnamoni*. This fungus can be spread by motor vehicles and livestock.

Victoria is a stark illustration of what can happen through the combined effects of habitat loss, human harassment and disease. It seems likely that we will soon witness the demise of this population unless decisive management action is taken in the near future.

The Worst Scenario

Loss of habitat poses the greatest problem for koalas. Young koalas are either prevented from dispersing from the established home ranges of their parents or must run the gauntlet of crossing open spaces to alternative habitat areas. During the breeding season koalas with amorous intent must dodge speeding motorists and avoid snapping dogs in order to seek mates. As urban society expands the habitat becomes more restricted and fragmented, dispersal of koalas ceases and overcrowding usually results. Stress associated with overcrowding or interference from human society, seems to predispose koalas to infection with *Chlamydia psittaci* and a combination of all factors inevitably leads to a local population decline.

Some of the by-products of human encroachment into forested areas act indirectly to the disadvantage of koalas. The accelerated water run-off resulting from our concrete and bitumen constructions may increase soil moisture and nutrient levels along drainage lines. This encourages the growth of non-eucalypt tree species so that eventually fewer koala food trees are present. Another indirect effect of urban sprawl on koalas results from the greater frequency of bushfires in urban fringing areas. The balance of tree species changes in forests which are regularly burnt and koala food trees are generally replaced with more fire-adapted species.

There is also a tendency for modern society to remove the shrub understorey of forests. This is done for aesthetic reasons, or as a fire preventative measure. Sometimes it is the result of grazing livestock. The loss of the shrub layer discourages insect-eating birds leaving the way open for voracious foliage-eating insects to compete with koalas for food. In the long-term, intense pressure from foliage-eating insects may result in 'dieback' of trees with a consequent reduction in the food trees available to koalas. The root rot fungus, *Phytophthora cinnamomi*, which can be spread by motor vehicles and livestock, also results in tree 'dieback' especially along roadsides and forest corridors.

The Future

If current trends in land clearing continue it is obvious that 'the worst scenario' will become reality for an increasing number of koala populations. Ultimately the future of koalas in some parts of their range may depend upon how many areas of suitable habitat can be set aside as nature conservation areas.

Some indication of the importance of nature conservation areas for koalas was provided by the National Koala Survey. In South Australia, ten of the thirty-two sites (31.2 per cent) where koalas were observed were in

150

151

152

153. Removing the shrub layer also has the effect of discouraging many of the insect-eating birds from the area and allowing foliage-eating insects, such as Christmas Beetles, to place pressure on the trees used by koalas for food.

153

national parks and nature reserves. The figure was slightly less in Victoria, where 151 of 696 (21.7 per cent) recorded sightings were at sites designated conservation areas. In New South Wales, 168 of the 697 (24.1 per cent) koala sightings were within areas controlled by the National Parks and Wildlife Service or Forestry Commission. Because various survey forms were used in Queensland, not all respondents were asked to provide details of land tenure. Of those that did, nineteen out of 554 (3.4 per cent) positive sightings of koalas were within nature conservation areas. These findings should only be seen as indicative as they are not directly comparable due to the different survey methods and forms used in each State.

The relatively good representations of koalas in nature conservation areas in South Australia and Victoria is partly a legacy of the translocation programs which have operated in these States for many years. The future of koalas in these southern States is relatively safe given that the relocation of koalas from overcrowded populations into larger forested areas of secure tenure continues. In this way the number of koalas within nature conservation areas continues to grow.

In time, wildlife managers in South Australia and Victoria may have difficulty in finding secure areas suitable for the relocation of surplus koalas. Lee and Martin (1988) have concluded that 'the pressing problem is to resolve what to do with koalas removed from overstocked areas'. They have suggested that '*Chlamydia* among koalas may be a blessing' as it has slowed the growth of a number of populations and served to delay the time when wildlife managers must answer the question 'what do we do when we have too many koalas'.

Optimism for the future of koalas may be appropriate in Victoria and South Australia but in New South Wales and Queensland there are strong indications that the situation is somewhat different. Although 24 per cent of koala sightings during the National Koala Survey in New South Wales were in nature conservation and Forestry Commission areas, these sightings were from a relatively few discrete populations. Koalas were present in only 29 of 289 (10 per cent) National Parks and Wildlife Service areas and 84 of 781 (10.8 per cent) New South Wales State Forests. These reserve areas are concentrated in the north-eastern quarter of the State.

The proportion of koalas in reserved areas in Queensland was less than in the other three States but this needs careful interpretation. As a general observation it seems that relatively few nature conservation areas in New South Wales and Queensland offer habitat suitable for koalas. Historically, the areas set aside for nature conservation were less valuable for primary production and, as a consequence, a high proportion of nature conservation areas do not include areas with the more fertile soils and the best koala food trees.

Additional nature conservation reserves which offer habitats best suited to the needs of koalas should be established in areas of prime value to rural, urban and tourism interests. The northern coast of New South Wales and southern coast of Queensland is where urban centres are expanding fastest to accommodate a thriving tourist industry. It is perhaps ironic that koalas are one of the Australian tourist industry's biggest drawcards, yet they are being threatened through relentless land clearing to establish new resorts and associated amenities. Unless key areas are preserved and development controlled to reduce further habitat fragmentation, it is unrealistic to expect anything but a progressive decline among many koala populations in this region.

Apart from the need to conserve large areas of pristine eucalypt forest suitable for koalas it is also important that the value of habitat corridors, or greenbelts, be recognised. It is not sufficient to simply set aside habitat islands to accommodate koalas. This is not a long-term solution to the problem of how to deal with koalas caught in the path of progress. Dispersal paths must be maintained or re-established so that contact with the general koala population, in larger forested areas, is not severed. Unless this occurs, more and more local populations will be lost and in the long-term the future of koalas on a regional basis may be threatened.

KOALAS TOMORROW

The future of koalas in modern Australia is a responsibility we all share. Each level of government has a role to play, as do all members of society. By recognising that loss of forest habitat poses the greatest potential threat to koalas, a greater responsibility for their future falls upon those people making decisions about how land is used and managed in the future. Elected representatives, government officials and private land owners, must carefully consider the implications of any decisions or actions which will affect, either directly or indirectly, areas occupied by koalas.

If the populations of koalas living on the edge of sprawling urban centres and areas of intensive primary production are to survive the time for action is now before the situation becomes impossible to rectify.

CONSERVING KOALAS: THE ROLES OF THE COMMONWEALTH GOVERNMENT

The Koala Conservation Program

The Commonwealth Government has responsibility for conserving wildlife and habitats of national or international significance. Koalas qualify in both respects and in recognition of this, the Koala Conservation Program was launched in June 1985.

The Koala Conservation Program has been administered by the Australian National Parks and Wildlife Service and financed by Commonwealth funds, public contributions and a donation of $200 000 from American Express International. Apart from coordinating and providing funds to assist the respective State wildlife authorities in undertaking the National Koala Survey financial support has been provided to a range of high priority research projects. Together with the findings of the national survey, the research projects have helped to clarify the current status of koalas in the wild and provided invaluable information which will assist in guiding the future conservation measures of the respective State wildlife authorities.

The development of formal strategies to protect koalas and their habitats must be undertaken at the State and Territory level as, under the Australian system of government, it is the States and Territories which have legal jurisdiction over land and how it is used. Through the Koala Conservation Program the Commonwealth Government has been assisting and encouraging the respective State governments to implement long-term measures to protect koalas. The preparation of detailed koala management plans is an important step towards achieving this goal and by 1990 the three major States where koalas are found should have operational koala management strategies.

The Koala Conservation Program has successfully achieved the major aims of examining the conservation status of koalas and identifying the factors which threaten their survival. The status of koalas needs to be kept under close scrutiny if this effort is not to be wasted.

155

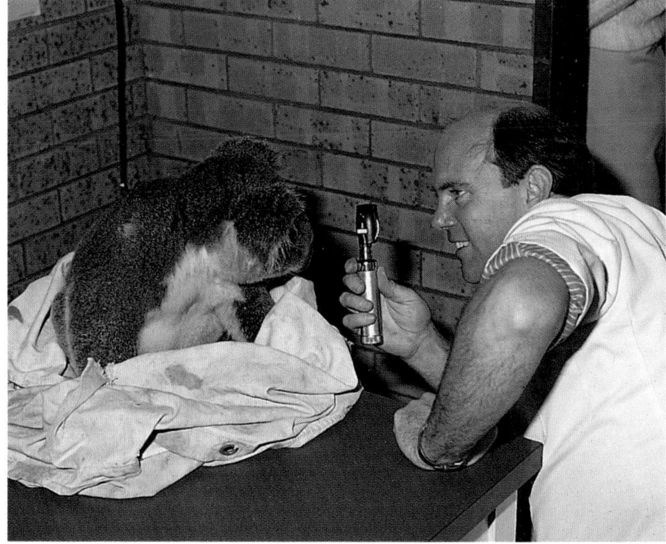

156

154. Through the Koala Conservation Program, financial support has been provided to a wide range of research projects.

155. Researchers and wildlife authorities catch a koala by using an extendable pole to place a noose around the koala's neck and then 'walking' the animal down the tree. Roger Martin of the Monash University research team demonstrates the technique.

156. Scientists have learnt a great deal about diseases caused by the bacterium *Chlamydia psittaci* and how these diseases can affect koalas.

The National Tree Program and 'Greening Australia'

Several years ago the Commonwealth Government recognised that injudicious forest clearing was beginning to adversely affect various sectors of the Australian community. On World Environment Day in 1982, the Prime Minister announced the establishment of the National Tree Program. The program, which continues today, is a coordinated attempt to change community attitudes towards tree conservation and reverse the trends which had seen nearly two thirds of the Australian tree cover removed in just 200 years. The National Tree Program, with continuing Commonwealth support, now forms part of the Greening Australia organisation, a non-government group with the same objectives.

There are encouraging signs that more and more people are getting involved in planting trees. As a result, several million trees have been planted in recent years and natural forest regeneration has been encouraged in many areas. While this is just a start, with increasing public awareness and support the momentum will hopefully continue to grow.

157. Through the National Tree Program and 'Greening Australia' several million trees have been planted over the past few years.

157

Farmers Leading the Way

The farming practices of yesteryear have been widely blamed for the loss and fragmentation of large proportions of our natural forests. Today the rural community is leading the way in protecting existing forests and in planting trees to rehabilitate other cleared areas. With the support of farmer organisations Farm Tree Groups have been established in all States.

Farm Tree Groups are mostly made up of farmers keen to promote,

KOALAS IN OVERSEAS ZOOS

The export of koalas for public display in overseas zoos has been a matter of debate. Opponents express concern for wild populations and the welfare of koalas kept in captivity. The proponents point to the benefits to be gained through increased public awareness of the biology and habitat needs of the species.

Through the provisions of the *Wildlife Protection (Regulation of Exports and Imports) Act* 1982, the Commonwealth Government has control over the export of koalas to overseas zoos. While Australia generally prohibits the export of live native animals for private or commercial purposes, the exchange of animals between publicly-owned zoos in Australia and their overseas counterparts is allowed. Exchanges are only permitted after it has been established that the overseas zoo is of high standing and has the expertise and proper facilities to ensure that animals sent from Australia are well cared for.

In the case of koalas, a detailed set of conditions governing their export was formulated by a group of zoo and nature conservation authorities experienced in the husbandry and maintenance of the species. These conditions, which are constantly under review, were first tabled in the Senate in 1980 and subsequently revised in 1984, 1986 and 1988.

The overseas zoos which have received koalas took several years making preparations to ensure they had adequate supplies of suitable eucalypt trees for food and high quality housing facilities. Veterinary staff and zookeepers were also trained to look after koalas. Before permits were issued to allow koalas to leave Australia, the recipient zoos were inspected. Special care was taken by the exporting Australian zoos to ensure that the koalas they sent were in very good health so that they would be able to withstand the trip.

Following these careful preparations, koalas have been sent to Japanese zoos in Tokyo, Nagoya, Kagoshima, Saitama, Yokohama and Awaji and to San Francisco Zoo in the United States of America.

Given the small number of animals sent overseas and the success of breeding at some institutions, the suggestion that the overseas export of koalas is depleting populations in Australia is not sustainable. Koalas chosen for export can only be taken from captive stock so wild populations are not affected.

On balance it seems the export of koalas has been important in enabling zoos to teach many people about koalas and in helping to develop an understanding of the significance of wildlife as well as an appreciation of the importance of its protection and conservation.

158

158. Sydney's Taronga Zoological Park plays an important role in increasing the general public's awareness of the special needs of species such as koalas.

through example, the importance of trees in rural areas. Increasingly, the farming community is realising the considerable benefits in retaining or planting trees on their properties. Trees help to reduce soil erosion and salinity levels, act as windbreaks, provide shade for livestock, and serve as shelter for wildlife, such as insect-eating birds and bats, which help to control insect pests.

159

159. National tree Program (NTP) pamphlets designed to encourage and help with tree planting activities.

160

160. Over the last twenty to thirty years the value of trees on farms has become more widely recognised. All States of Australia now have farmer organisations promoting and cooperating in tree planting programs.

FUND-RAISING FOR KOALA RESEARCH

In October 1985 the Australian Wildlife Fund Limited was launched by the then Minister for Arts, Heritage and Environment, Mr Barry Cohen. Established principally to raise funds for the research projects of the Australian National Parks and Wildlife Service the initiative was also seen as an opportunity to increase public awareness of wildlife conservation issues.

In 1986, as its first fund-raising activity, the Australian Wildife Fund produced a set of koala research stamps. Although not intended for postage purposes, similar 'cinderella' stamps, as they are known, have proven very popular with stamp enthusiasts in other parts of the world for many years. They have been successful in generating funds for wildlife conservation and management.

Further releases of koala research stamps were made in 1987 and 1989. Proceeds from the sale of these stamps have been provided to the Australian National Parks and Wildlife Service to assist with the funding of further koala research projects. It is hoped that as the fund-raising efforts of the Australian Wildlife Fund gain momentum financial support for more wildlife research projects will be available.

For further information about koala research stamps and other associated products contact:

The Australian Wildlife Fund
Post Office Box 653
North Sydney NSW 2059

KOALA RESEARCH AUSTRALIAN NATIONAL PARKS AND WILDLIFE SERVICE 50¢
KOALA RESEARCH AUSTRALIAN NATIONAL PARKS AND WILDLIFE SERVICE $1
KOALA RESEARCH AUSTRALIAN NATIONAL PARKS AND WILDLIFE SERVICE $1.50
KOALA RESEARCH AUSTRALIAN NATIONAL PARKS AND WILDLIFE SERVICE $2

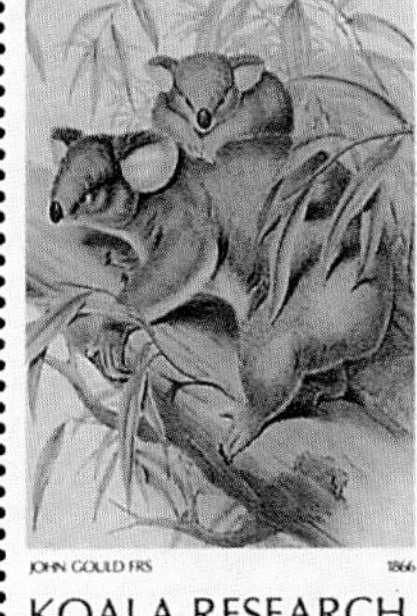

KOALA RESEARCH AUSTRALIAN NATIONAL PARKS AND WILDLIFE SERVICE $2
KOALA RESEARCH AUSTRALIAN NATIONAL PARKS AND WILDLIFE SERVICE $3

161

161. The Australian Wildlife Fund was launched in 1985 to raise funds for the wildlife research programs of the Australian National Parks and Wildlife Service. The first initiative was to support the Koala Conservation Program with three issues of koala research stamps

CONSERVING KOALAS: THE ROLES OF STATE AND LOCAL GOVERNMENT

162. While there is no simple solution to balancing conservation objectives with the need for a viable timber industry, the move towards eucalypt plantations is one option worthy of consideration.

163. Under the Australian system of government the States and Territories have legal jurisdiction over land and how it is used. However, local governments also have a responsibility to preserve forest habitats that border towns and cities.

The continuing loss of habitats suitable for koalas can only be halted if State and local governments recognise their responsibilities and cooperate in preserving areas of natural forest. State governments already have in place the legislation necessary to protect key areas of koala habitat. Local governments, through their many contacts with the general public, are ideally placed to monitor land use and encourage habitat conservation.

Formal koala management strategies are currently being prepared in Victoria, New South Wales and Queensland. These will identify locations where koala conservation is of greatest concern and establish guidelines by which these populations can be more effectively managed and protected. Populations under threat must be closely monitored to enable State government agencies and local governments to take suitable corrective actions where necessary.

The declaration of additional protected areas which include good koala habitats is an important initiative for State governments to consider, especially in New South Wales and Queensland where relatively few nature conservation reserves contain koala populations. Sufficient is known of the habitat needs of koalas to identify and set aside ideal areas specifically for the conservation of this unique species.

State governments should also look at areas currently designated as Crown Land to see if they contain suitable habitat for koalas. In rural areas, the now little used, travelling stock routes, sometimes form convenient corridors along which koalas and other wildlife can move. Clearing these routes only serves to further isolate forest remnants. Consideration should also be given to retaining suitable Crown Reserves near towns and cities. This is not to say that all property development should cease but rather that through thoughtful land management, prime areas of forest habitat could be preserved and areas less important to wildlife used for housing estates, highways or easements for essential services such as electricity, water, gas and telephone.

Local governments can help in land planning and negotiating with property developers to provide a suitable compromise in protecting wildlife habitat and accommodating the needs of expanding urban centres. State governments must lend support to the efforts of local government by strengthening environment protection legislation and being prepared to provide expert input into land development decisions.

Tangible support for tree-planting programs should be forthcoming from all levels of government. Community groups and farmers' associations have already shown their will-

162

163

164. In many areas tree planting programs are vital to the future of koalas and the host of other species which share their leafy domain.

164

ingness to be involved in these programs. It would be irresponsible of governments, at all levels, not to encourage and foster these enthusiastic and committed groups to continue with their tree-planting programs.

Governments also need to consider programs which give tax remission to landholders who conserve or add to forested areas. The Commonwealth Government abolished tax deductibility for the costs of land clearing several years ago and, at the same time, introduced tax incentives for tree and conservation activities. Similar tax incentives at the State and local government levels could assist greatly in curtailing the relentless removal of precious forest habitats.

The establishment of private nature reserves is another way wildlife habitats can be preserved. Some States already have provision for such arrangements, whereby landholders voluntarily enter into an agreement to maintain their land as wildlife refuge. Given that a large proportion of our koalas live on privately owned land, this initiative should be given high priority by the respective State governments.

A further positive step towards habitat conservation is that of Tree Preservation Orders which have now been introduced by many local shires and city councils. Tree

'BUY HERE AND LIVE AMONG THE KOALAS'

'Buy here and live among the koalas' was an advertising sign once used to entice potential homeowners to purchase land on Sydney's Barrenjoey Headland. Against all odds, koalas are still present in the nearby Avalon area today despite being completely surrounded by urban development.

Substantial credit for the koalas still being found in the Avalon area must go to the local Preservation Trust. This group of concerned citizens first took interest in the welfare of the local koalas thirty years ago. Koalas were then present throughout large areas of bushland on Bilgola Plateau extending from Newport to Palm Beach. With urban development the bushland areas have been gradually eroded until only the tiny 'islands' found in Avalon remain. Miraculously koalas have survived there. Major initiatives of the Avalon Preservation Trust have been to educate the public about koalas and to keep a watchful eye on developers to ensure the precious koala food trees are protected.

The Avalon Preservation Trust were instrumental in having a Tree Preservation Order introduced into Warringah Shire and this combined with tree planting helped to maintain sufficient food trees for the local koalas. Because the Preservation Trust has developed good relations with the Warringah Shire Council, its members inspect any new subdivision sites and make recommendations to minimise tree loss and habitat damage.

Peter Smith, the Environmental Officer with Warringah Shire Council, suggests that the survival of the Avalon koala colony has been assisted by several factors. The size of housing blocks in the area was initially large and some had tree covenants placed on them at sale by an environmentally aware real estate agent. Avalon also has a high proportion of bushland reserves and the topography of the area is such that the road system does not permit high speed traffic.

Even though the Avalon koala colony is apparently coping, the long-term future of the population must be in doubt unless steps are taken to prevent further loss of food trees. Old trees need to be replaced with younger ones to provide for the future. A major concern is that the koala population has probably been isolated from the general population for around forty years. Without the introduction of new animals inbreeding must inevitably result.

The fact that the Avalon koalas have survived in their remnant habitat is a tribute to the tenacity of the species and the dedication of concerned citizens. Unfortunately, this case is more likely the exception than the rule.

166

165

165 & 166. A sea of urbanisation now surrounds the forest remnants at Avalon, on Sydney's north shore, where a small koala colony miraculously survives.

167. Tree Preservation Orders restricting landowners from removing trees above a certain size are an important local government mechanism for protecting wildlife habitat.

168. Koala up a tree

169. A road sign to warn motorists that koalas are about.

Preservation Orders should be accompanied by public education programs so that tree retention is a voluntary action rather than an obligation imposed by council.

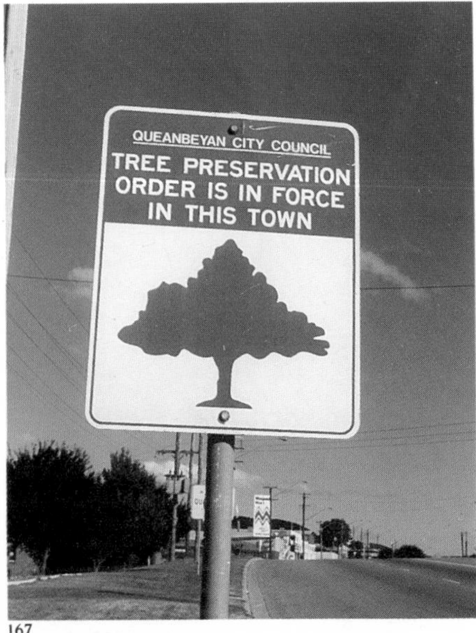

167

The Importance of Public Awareness

Although several organisations currently distribute information intended to encourage nature conservation, there is a need to escalate these programs. Consideration for the environment only comes with awareness and an appreciation of the implications of haphazard land use.

Information packages, explaining the benefits of habitat corridors and preserving forests, should be developed and targeted specifically at people such as town planners and land developers. Local councils need to be advised of measures they can take to increase public concern for wildlife and involvement in conservation measures. The escalating koala mortality from motor vehicles and domestic dogs can be drastically reduced through public awareness campaigns and by road signs in key koala areas warning motorists to drive cautiously. The intricacies of our natural environment should be a major emphasis of schooling, particularly in rural areas where todays schoolchildren may one day be making decisions which will directly affect the delicate balance of nature.

168

169

PLANTING TREES FOR KOALAS

Tree planting programs hold the key to conserving koalas. Whether it be through the re-establishment of once forested areas or the planting of tree corridors linking larger areas of existing forest, koalas will certainly benefit.

Most trees take about fifteen years before they are suitable as food trees for koalas so the time for action is now, especially in those areas where koalas are immediately threatened through habitat loss.

Koalas prefer different tree species in different parts of their range. Tree planting programs must therefore take account of local and regional variations in food choices. The results of the National Koala Survey provide an indication of koala food tree preferences on a State-by-State basis.

South Australia

Manna Gum	*Eucalyptus viminalis*
South Australian Blue Gum	*Eucalyptus leucoxylon*
River Red Gum	*Eucalyptus camaldulensis*

Victoria

Manna Gum	*Eucalyptus viminalis*
Messmate Stringybark	*Eucalyptus obliqua*
Swamp Gum	*Eucalyptus ovata*
Narrow-leaved Peppermint	*Eucalyptus radiata*
Tasmanian Blue Gum	*Eucalyptus globulus*
River Red Gum	*Eucalyptus camaldulensis*

New South Wales

Sydney Blue Gum	*Eucalyptus saligna*
Grey Gum	*Eucalyptus punctata*
Flooded Gum	*Eucalyptus grandis*
Blackbutt	*Eucalyptus pilularis*
Tallow-wood	*Eucalyptus microcorys*
Small-fruited Grey Gum	*Eucalyptus propinqua*

Queensland

Forest Red Gum	*Eucalyptus tereticornis*
Narrow-leaved Red Ironbark	*Eucalyptus crebra*
Poplar Box	*Eucalyptus populnea*
River Red Gum	*Eucalyptus camaldulensis*
Spotted Gum	*Eucalyptus maculata*
Small-fruited Grey Gum	*Eucalyptus propinqua*

CONSERVING KOALAS: THE GENERAL COMMUNITY

More than 50 per cent of koalas live on privately-owned land and so their future is largely dependant on the attitude and actions of landowners. Apart from protecting existing forested areas, private landholders have the opportunity to replant trees in areas where koalas were once found. Hobby farmers and those living on larger semi-rural blocks have a unique opportunity, either individually or in cooperative groups, to maintain or establish habitat corridors linking larger areas of forest.

Tree-planting programs are usually met with such enthusiasm that the funding allocation from government bodies is insufficient to meet the requests from community groups. This is an area in which the private sector should get more involved so as to harness this enthusiasm and energy. From the earliest days of commercial enterprise in Australia koalas have been lending their name and image to promote products and companies. Very few companies have seen fit to repay the debt by sponsoring conservation-directed programs. The infusion of relatively minor sums into tree planting programs would not only benefit koalas but also generate substantial goodwill for the companies concerned.

The restoration of forest habitats is not simply a matter of digging holes and placing trees into them. Tree species selection for each region has to be given careful consideration and best results are obtained when there is follow-up care to ensure losses due to frost, drought, fire and livestock are minimised. A long term community involvement is essential.

Being Koala-conscious

People living near areas occupied by koalas have special responsibilities and need to be koala-conscious at all times. Dogs should be restrained at night and extreme caution shown when driving vehicles, especially during the warmer months of the breeding season when koalas are more mobile and more likely to be on the roads. In suburbs adjoining natural forests, backyards should be made more koala-friendly. Garden

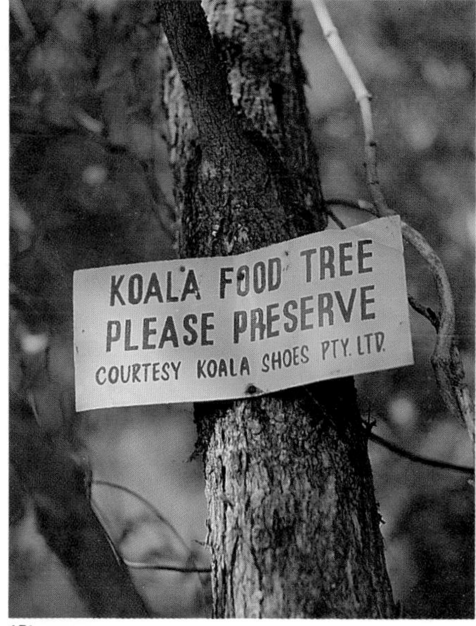

171

170. Private landowners must assume a greater responsibility for protecting and replacing wildlife habitats. More than 50 per cent of koalas live on land which is privately owned.

171. Koalas have been used for years to promote products and private businesses.

172. Community groups help sick, injured and orphaned koalas.

170

172

173. Children are told of the importance of protecting and caring for wildlife.

sheds, which may contain hazards such as toxic chemicals and paints, should be kept closed at night and a timber climbing pole may allow koalas to scale otherwise impassable fences or extricate themselves from in-ground swimming pools.

Sick, injured and disoriented koalas will inevitably come into contact with people. Koalas in obvious need of veterinary care or relocation should be reported to the local office of the National Parks and Wildlife Service, the Royal Society for the Prevention of Cruelty to Animals or any community group specialising in the care and rehabilitation of koalas. Despite their cuddly image, koalas should not be handled by the inexperienced as they can inflict nasty injuries if handled incorrectly.

Lost koalas should not be relocated without first consulting the appropriate wildlife authorities. A special permit may be required and the innocent relocation of a koala infected with *Chlamydia psittaci* into a population free of infection may result in the rapid decline of the previously unexposed and highly susceptible koala community.

A number of community groups operate in areas where koalas are most immediately threatened by urban expansion. They help sick and injured koalas, rescue lost, disoriented or harassed koalas, undertake tree-planting programs and generally encourage awareness of the conservation issues facing koalas.

173

A COMMUNITY SHOWS THEY CARE FOR KOALAS

Since 1973 a few concerned citizens in the New South Wales town of Port Macquarie have been fighting to preserve their dwindling koala population.

Scenic Port Macquarie on the northern coast of New South Wales was once prime koala habitat. Since the establishment of a penal settlement there in 1821, vast areas of forest have gradually been cleared to make way for a thriving tourist centre.

Port Macquarie provides a graphic example of how urban expansion inevitably threatens the koalas living on the adjoining land. On the brighter side, the actions of the Koala Preservation Society of New South Wales are an example of how a community can work together to help conserve wildlife in an urban environment.

In 1973 the population of Port Macquarie was 11 000 and it was obvious that the town was in the midst of a major population explosion which would inevitably destroy the remaining koala habitat. Concerned for the future of koalas in and around the town, a small group of citizens, with the support of the New South Wales National Parks and Wildlife Service and the Hastings Municipal Council formed the Port Macquarie Koala Preservation Society (later to become the Koala Preservation Society of New South Wales). The aim of the society was to engender a sense of cooperation in the community to protect koalas and to promote forward planning to conserve koala habitat.

The population of Port Macquarie is now approaching 27 000 and nearly 700 000 tourists visit the town annually. To those living in or visiting the town, the results of the efforts of the Koala Preservation Society are clear. Habitat corridors now run through the town linking patches of coastal forest with larger inland forest reserves. These greenways allow young koalas to disperse and have meant that a viable koala population continues to exist in downtown Port Macquarie.

Working with the Hastings Municipal Council in 1973 the Society identified four key habitat corridors.

175

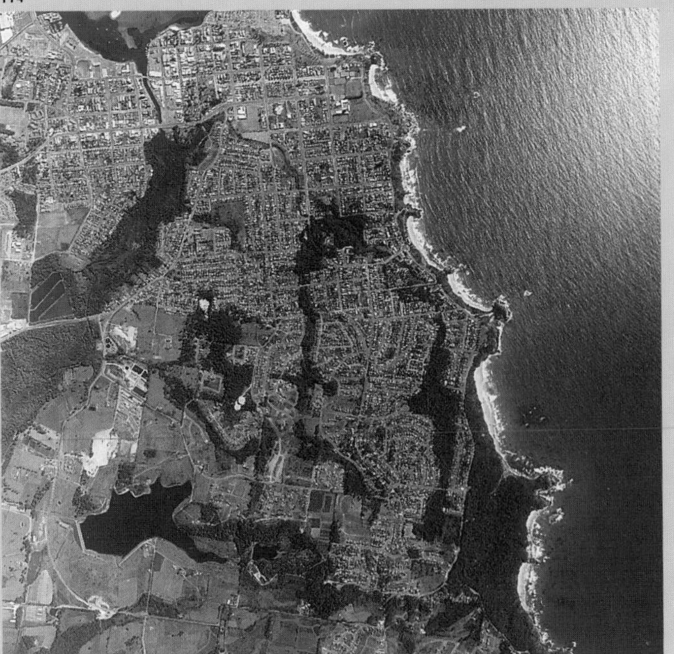

174

174. The Koala Preservation Society of New South Wales has been instrumental in establishing a network of wildlife corridors linking larger forest reserves in the expanding tourist centre of Port Macquarie on the northern New South Wales coast.

175. Thanks to the people of Port Macquarie koalas are still present in the area today.

176. The Koala Preservation Society has also established a koala hospital and study centre at Port Macquarie.

Many hours were spent watching the local koalas so as to determine which were their preferred tree species. Local forestry officers showed the society members how to collect and propagate the seeds of these eucalypts. Fifteen years later and with the assistance of many service clubs, cubs, scouts, brownies and members of the public, over 11 000 trees have been planted.

On one notable morning when a cyclone brought 50 millimetres of rain, fifty dedicated people planted 1 000 trees in just two hours. Although some trees were later found to be planted upside down in the haste, the operation was a success and helped to establish the network of greenways now evident in Port Macquarie.

Although the habitat corridors have proved successful and koalas and other wildlife are still present in Port Macquarie, members of the Koala Preservation Society are concerned for the long term future of the koalas. Urban sprawl continues to push the outlying koala populations further away from those in Port Macquarie and despite the habitat corridors, substantial numbers of koalas are still killed on the roads.

To cope with the increasing number of sick and injured koalas, the Koala Preservation Society has established a Koala Hospital and Study Centre at Port Macquarie using funds donated by the Sun-Herald Koala Fund, the Australian National Parks and Wildlife Service and the Hastings Municipal Council. At the hospital, sick and injured koalas are treated and rehabilitated and orphaned cubs are reared and then released into the wild.

The Koala Hospital incorporates a study centre designed to improve the public perception and understanding of koalas. There are static displays explaining aspects of koala biology and their habitat needs. Educational material has been produced, some of which is designed for use by teachers.

The achievements of the Koala Preservation Society at Port Macquarie demonstrate the sorts of initiatives needed to ensure that rapidly expanding towns and cities do not completely engulf local koala populations.

Well-known country singer and songwriter John Williamson reflects these sentiments in his song 'Goodbye Blinky Bill'. Concern for the future of koalas has been shown by the public support for the song which has seen well over $100 000 in proceeds go to help the Koala Preservation Society of New South Wales.

176

178

177

179

177. Sick and injured koalas and orphaned cubs are treated and rehabilitated and then released into the wild.

178. The Koala Preservation Society raises money at local shows and fetes.

179. Well-known country singer and song writer, John Williamson, donated the proceeds from his song 'Goodbye Blinky Bill' to the Koala Preservation Society.

KOALAS: THE LITTLE AUSTRALIANS WE'D ALL HATE TO LOSE

On a tiny watercolour painting adorning the wall of a classroom is a koala clinging to a street light with the words written by a child, 'koalas can't live on lamp posts'. The young artist could not have known when painting this work how accurately these words describe the threat to koalas in modern Australia. The expansive forests where koalas once lived, in their inimitable and unassuming way, have largely gone and those which remain are rapidly disappearing to make way for the needs of human society.

Since arriving in Australia, European settlement has removed nearly two-thirds of our tree cover and has been responsible for the extinction of at least ninety-seven plant, two bird and eighteen mammal species. Today, a further 209 plants, thirty-seven birds, six amphibians, nine reptiles and forty-one mammals are listed as endangered species.

Current evidence suggests that koalas are not in imminent danger of extinction but the warning signs are there. The threats have been recognised and ignorance can no longer be an excuse for driving species towards extinction.

Koalas have shown their resilience by recovering from the past depredations of hunters and diseases, but they are inextricably bound to eucalypt forests. If the forests go, koalas will certainly go with them. Superficially, the answer to the problem appears quite simple. Stop cutting down and bull-dozing trees. In reality the problem is not readily solved as the solution lies in changing social attitudes. Indifference towards the environment needs to be replaced by an understanding that the relentless clearing of forests cannot continue as it has for the past

180. The words of a child aptly describe the situation of koalas in modern Australia.

181. A koala caught up a telegraph pole.

182. Habitat loss poses the greatest threat to koalas in Australia today.

180

181

200 years, without ultimately costing Australia dearly.

The eucalypt forests are a limited resource, arguably our most precious. The key to their preservation lies in generating interest in the environment and an awareness of the way natural ecosystems and food webs operate and serve to sustain the delicate balance of nature. A balance humankind seems hellbent on destroying. Despite a general public perception to the contrary, conservation need not lead to confrontation. A groundswell of public concern will have far greater benefits for environment conservation in the long-term. As an affluent, resourceful and technologically-advanced nation it seems inexplicable that Australia should drive species to the brink of extinction before the hard decisions can be made.

The major responsibility for protecting the environment has for too long fallen upon governments. The plight of koalas highlights the vital importance of self-imposed nature conservation measures by private landowners. If past land management practices had been more sensitive to the environment, there would be no need for governments to set aside unique areas for future generations.

The rural community has realised through experience that overzealous land clearing, in order to maximise grazing and pastoral areas, ultimately has far-reaching repercussions which ironically result in reduced productivity. The value of trees on farms is increasingly being recognised and management practices in rural areas are gradually moving towards a more acceptable balance between workable land and natural habitat. Through this approach wildlife benefits and so too does the primary producer.

It remains to be seen whether the tourism and manufacturing industries can also learn to treat the natural environment with suitable respect. As vital as the tourist dollar and dynamic industrial enterprises are to sustaining Australia's current standard of living, it would be foolhardy and shortsighted if this were at the expense of the environment.

Unfortunately, suburban sprawl now poses a serious long-term threat to wildlife, and notably koalas. The forests which harbour much of our wildlife are also where our major cities and tourist areas have developed. There is no simple solution to the problem of balancing the inevitable urban expansion of a growing nation with the conservation

182

183. One of the 'little Australians we'd all hate to lose'.

183

of forests and the wildlife they support.

The establishment of national parks and nature reserves is one positive step governments can take but this is only part of the answer. Unless developers adopt more environmentally sensitive attitudes, the future of a large number of species is without doubt threatened.

With the removal of such a large proportion of our forests the conservation of many species, including koalas, is now reliant upon protecting the good quality habitats which remain. By necessity, habitat conservation now needs to be more about quality than quantity. To achieve this, compromises will have to be made and in some instances short-term profit sacrificed in favour of long-term environmental benefits. This conflict of interest is not new to Australia, nor indeed the rest of the world, but over the past decade a growing public concern has seen private developers and government authorities alike placed more and more under the microscope in their dealings with the environment. This is a trend which must be encouraged as it can only benefit the causes of wildlife conservation and environment protection in general.

It should be a sobering thought that if current trends of land clearing are allowed to continue, it will not only be koalas which suffer. Despite the reputation of the seething rainforests as species-rich environments, the eucalypt forests of the fertile plains and alluvial river flats are also vital to the survival of vast communities of species. Koalas are among these species and because they are blessed with a pleasing public image concern for their future has been inordinately vocal. However, it would be irresponsible to ignore the many others species which share the forests and trees with koalas. Although less appealing or less well known to the public, they are equally precious. Greater effort should be directed at conserving forest habitats for all species, not just koalas.

The species which make up the world of koalas are many and varied. The possums and gliders and many insectivorous bats seek refuge in and feed from and around the eucalypts preferred by koalas. Owls and parrots and a host of tiny insect-eating birds are also concentrated in these areas. Many migrating honey-eaters move along rivers and streams, feeding and resting in the trees of the banks and adjoining woodlands. Lyrebirds, bowerbirds and echidnas scurry amongst the leaf litter on the forest floor and platypus burrow into the banks secured by the roots of river red gums. Snakes, goannas and water dragons bask in the sun as wombats snooze in their burrows and wallabies cautiously graze on grass-shoots.

This, but superficial, look at the complex ecosystem which koalas sit quietly representing, should serve to remind us that conservation needs to be about habitats and not just individual species. If habitats are protected so too will be their occupants.

For 200 years the eucalypt forests of Australia have been seen either as impediments to development or a resource to be exploited. The threat to koalas which is now looming on the horizon should be recognised for what it is — a warning that all is not well in the 'lucky country'. If we would jeopardise the future of such a precious species our priorities are clearly wrong. The future of koalas rests firmly in our hands. No one would argue that they really are the little Australians we'd all hate to lose.

BIBLIOGRAPHY

Archer, M. and Hand, S. 1987, 'Evolutionary Considerations', in *Koala, Australia's Endearing Marsupial*, ed L. Cronin, Reed Books Pty Ltd, Frenchs Forest, NSW., pp.79–106

Backhouse, T.C. and Bolliger, A. 1961, 'Morbidity and Mortality in the Koala (*Phascolarctos cinereus*), *Australian Journal of Zoology*, Vol. 9, pp. 24–37

Backhouse, G. and Crouch, A. 1988, 'Koala Management in the Western Port Region of Victoria', presented to the Third Symposium on the Biology of the Koala, Melbourne, February 1988

Barallier, F. 1802, quoted in 'The Early History of the Koala' by T. Iredale and G.P. Whitley, *Victorian Naturalist*, Vol. 51,

Barallier, F. 1803, The *Sydney Gazette*, 21 August 1803

Bray, T.L. 1959, 'Some Recent Events of Interest', *Wildlife Service*, Vol. 1(6), p.15.

Brough Smyth, R. 1876, *The Aborigines of Victoria*, Vol. 1, John Currey, O'Neil, Melbourne

Brown, A.S. and Carrick, F.N. 1985, 'Koala Disease Breakthrough', *Australian Natural History*, Vol. 21. pp. 314–7

Brown, A.S. and Grice, R.G. 1984, 'Isolation of *Chlamydia psittaci* from Koalas (*Phascolarctos cinereus*)', *Australian Veterinary Journal*, Vol. 50, pp. 82–3

Brown, A.S., Carrick, F.N., Gordon, G. and Reynolds, K. 1984, 'Diagnosis and Epidemiology of an Infertility Disease in the Female Koala', *Veterinary Radiology*, Vol. 25, pp. 242–8

Bullock, W. 1814, A *Companion to the London Museum and Pantherion* (17th Edition)

Campbell, W.D. 1899, 'Aboriginal Carvings of Port Jackson and Broken Bay', *Memoirs of the Geological Survey of New South Wales, Ethnological Series No. 1*

Campbell, P., Prentice, R. and McRae, P. 1979, 'Report on the 1977 Koala Survey', *Wildlife in Australia*, Vol. 16(1), pp. 2–6.

Canfield, P. 1987, 'A Study of Koala Deaths', *Australian Science Magazine*, Vol. 4, pp. 24–5.

Clegg, J. 1988, 'Berowra Waters Koala Engravings', paper presented to Koala Summit, New South Wales National Parks and Wildlife Service, Sydney, November 1988

Cockram, F.A. and Jackson, A.R.B. 1974, 'Isolation of *Chlamydia* from Cases of Keratoconjunctivitis in Koalas', *Australian Veterinary Journal*, Vol. 50, pp. 82–3

Cockram, F.A. and Jackson, A.R.B. 1981, 'Keratoconjunctivitis of the Koala, *Phascolarctos cinereus*, caused by *Chlamydia psittaci*', *Journal of Wildlife Diseases*, Vol. 17, pp. 497–504

Congreve, P. and Betts, T.J. 1978, 'Eucalyptus Plantations and Preferences as Food for a Colony of Koalas in Western Australia', in *The Koala, Proceedings of the Taronga Symposium*, ed. T.J.Bergin, John Sands Pty Ltd, Artarmon, NSW

Cork, S. 1987, 'Form and Function in the Koala', in *Koala, Australia's Endearing Marsupial*, ed. L. Cronin, Reed Books Pty Ltd, Frenchs Forest, NSW, pp. 31–55

Coutts, P.F. 1970, 'The Archaeology of Wilson's Promontory', *Australian Aboriginal Studies No. 7*, Australian Institute of Aboriginal Studies, Canberra

Coverdale, T.J. 1920, 'The Scrub', in *The Land of the Lyrebird*, The South Gippsland Pioneers Association, Gordon and Gotch, Melbourne, pp. 16–23

Cunningham, A. 1818, quoted in 'The Early History of the Koala' by T. Ireland and G.P. Whitley, *Victorian Naturalist*, Vol. 51, Melbourne

Cuvier, G. 1817, *Regne Animal*, Australian Museum Library

Dickens, R.K. 1978, 'The Koala in Health and Disease', in Proceedings No. 36 of Course for Veterinarians, Fauna Part B, the Postgraduate Committee in Veterinary Science, University of Sydney, Sydney, pp. 105–17

Dixon, S. 1920, *The Full Story of Flinders Chase, Kangaroo Island, South Australia: a New Holiday and Health Resort for South Australians and Visitors from other Parts with Photographs and Large Maps*, Hussey and Gillingham, Adelaide

Eberhard, I.H. 1972, 'Ecology of the Koala, *Phascolarctos cinereus* (Goldfuss) on Flinders Chase, Kangaroo Island', Ph.D. Thesis, University of Adelaide

Eberhard, I.H. 1978, 'Ecology of the Koala, *Phascolarctos cinereus* (Goldfuss), Marsupialia: Phascolarctidae, in Australia', in *The Ecology of Arboreal Folivores*, ed G.G. Montgomery, Smithsonian Institution Press, Washington DC, pp. 315–28

Eberhard, I.H. and Schulz, L. 1973, 'A Survey of the Vertebrate Fauna of the Cotter River Catchment', Australian Capital Territory, Department of the Capital Territory Conservation Memorandum, Canberra

Fountain, P. 1907, *Rambles of an Australian Naturalist*, John Murray, London

Gall, B.C. 1978, 'Koala Research, Tucki Nature Reserve', in *The Koala, Proceedings of the Taronga Symposium*, ed T.J. Bergin, John Sands Pty Ltd, Frenchs Forest, NSW, pp. 116–24

Gall, B.C. and Rohan–Jones, W. 1978, 'Koala Survey', Parks and Wildlife 2, pp. 64–7

Girjes, A.A., Hugall, A.F., Timms, P and Lavin, M.F. 1988, 'Two Distinct Forms of *Chlamydia psittaci* Associated with Disease and Infertility in *Phascolarctos cinereus* (Koala)', *Infection and Immunology* (in press)

Glauert, L. 1910, 'The Mammoth Cave', *Records of the Western Australian Museum*, pp.11–36

Gordon, G., Brown, A.S. and Pulsford, T. 1988, 'A Koala (*Phascolartos cinereus*) Population Crash during Drought and Heatwave Conditons in South–western Queensland', *Australian Journal of Ecology*, Vol. 13, pp. 451–61

Gordon, G. and McGreevy, D.G. 1978, 'The Status of the Koala in Queensland', in *The Koala, Proceedings of the Taronga Symposium*, ed T.J. Bergin, John Sands Pty Ltd, Frenchs Forest, NSW, pp.125–31

Gordon, G., McGreevy, D.G. and Lawrie, B.C. 1989, paper presented to the Third Symposium on the Biology of the Koala, Melbourne, February, 1988

Gould, J. 1863, *The Mammals of Australia*, Taylor and Francis, London

Govatt, W.R. 1836, 'Sketches of New South Wales No. XIV on the animals called "monkeys" in New South Wales', the *Saturday Magazine*, Vol. 9, pp. 249–50

Handasyde, K.A. 1986, 'Factors Affecting Reproduction in the Female Koala (*Phascolarctos cinereus*)', Ph.D. Thesis, Monash University, Clayton, Victoria

Hardy, A.D. 1906, 'Excursion to Wilson's Promontory', *Victorian Naturalist*, Vol. 22, pp. 191–7

Haydon, G.H. 1846, *Five Years Experience in Australia*, Felix. Hamilton Adams and Co., London

Hindell, M.A. 1984, 'The Feeding Ecology of the Koala, *Phascolarctos cinereus*, in a Mixed *Eucalyptus* Forest', M.Sc. Thesis, Monash University, Clayton, Victoria

Home, E. 1808, 'An Account of Some Peculiarities in the Anatomical Structure of the Wombat', *Philosophical Transactions of the Royal Society of London*, pp. 308–12

Iredale, T. and Whitley G.P. 1934, 'The Early History of the Koala', *Victorian Naturalist*, Vol. 51, pp. 62–72

Kikkawa, J.A. and Walter, M. 1968, 'Report on the Koala Survey', *Wildlife in Australia*, Vol. 6, pp. 100–3

Lee, A.K. and Martin, R.W. 1988, *The Koala: a Natural History*, New South Wales University Press, Kensington, NSW

Lee, A.K., Martin, R.W. and Handasyde, K.A. 1988, 'Research into Chlamydiosis in Natural Populations of the Koala in Victoria, Phase III', Report to the Australian National Parks and Wildlife Service, Canberra

Lewis, F. 1934, 'The Koala in Victoria', *Victorian Naturalist*, Vol. 51, pp.73–6

Lewis, F. 1954, 'The Rehabilitation of the Koala in Victoria', *Victorian Naturalist*, Vol. 70, pp. 197–200

Lydekker, R. 1894, A *Handbook to the Marsupialia and Monotremata*, W.H. Allen and Co. Ltd, London

Marshall, A.J. 1966, 'On the Disadvantages of Wearing Fur', in *The Great Extermination*, Melbourne, pp. 26–33

Martin, R.W. 1981, 'Age-specific Fertility in Three Populations of the Koala, *Phascolarctos cinereus* (Goldfuss) in Victoria' *Australian Wildlife Research*, Vol. 8, pp.275–83

Martin, R.W. 1983, 'Food Preference, Defoliation and Population Decline in a Population of the Koala, *Phascolarctos cinereus*, at Walkerville, Victoria' M.Sc. Thesis, Monash University, Clayton, Victoria

Martin, R.W., Handasyde, K. and Lee, A. 1987, 'Is the Koala Endangered?' *Australian Science Magazine*, Issue 4, pp. 26–30

McCarthy, F.D. 1956, 'Rock Engravings of the Sydney-Hawkesbury District Part 1: Flat Rocks Ridge: a Daruk Ceremonial Ground' *Records of the Australian Museum*, Vol. XXIV, No. 5, pp.37–58

McCarthy, F.D. 1961, 'A Remarkable Ritual Gallery of the Cave Paintings in Eastern New South Wales', *Records of the Australian Museum*, Vol. XXV, No.7, pp.115–20

McColl, K.A., Martin,R.W.,Gleeson, L.J., Handasyde, K.A. and Lee,A.K. 1984, '*Chlamydia* Infection and Infertility in the Female Koala (*Phascolarctos cinereus*)', *Veterinary Records*, Vol. 115, p. 655

McQueen, H. 1978, *Social Sketches of Australia* 1888–1975, Penguin Books, Victoria

Mitchell, P. 1988, 'Social Organisation of the Koala, *Phascolarctos cinereus*' Ph.D. Thesis, Monash University, Clayton Victoria

Murray, D.W. 1988, 'The Distribution of the Koala, *Phascolarctos cinereus* (Goldfuss) Predicted by Computer Analysis of Selected Environmental Factors', paper presented to the Third Symposium on the Biology of the Koala, Melbourne, February 1988

Nagy, K.A. and Martin, R.W. 1985, 'Field Metabolic Rate, Water Flux, Food Consumption and Time Budget of Koalas, *Phascolarctos cinereus* (Marsupialia: Phascolarctidae) in Victoria', *Australian Journal of Zoology*, Vol. 33, pp.655–65

Obendorf, D.L. 1983, 'Causes of Mortality and Morbidity in Wild Koalas, *Phascolarctos cinereus* (Goldfuss) in Victoria, Australia, *Journal of Wildlife Diseases*, Vol. 19, pp.123–31

O'Donoghue, C.H. 1916, 'On the Corpora Lutea and Interstitial Tissue of the Ovary in the Marsupialia', *Quarterly Journal of Microscopic Science*, Vol. 61, pp.433–73

Pahl, L.I. and Hume, I.D. 1988, 'Preferences for the *Eucalyptus* Species of the New England Tablelands and Initial Development of an Artificial Diet for Koalas' paper presented to the Third Symposium on the Biology of the Koala, Melbourne, February 1988

Parris, H.S. 1948, 'Koalas on the lower Goulburn', *Victorian Naturalist*, Vol. 64, pp.192-3,

Pelsaert, F. 1628, quoted in 'Introduction to the Marsupials' by S. Cork, from *Koala; Australia's Endearing Marsupial*, ed L. Cronin, Reed Books Pty Ltd, Frenchs Forest, NSW, p.9

Perry, G. 1810, *Arcana*, James Stratford, London

Pratt, A. 1937, *The Call of the Koala*, Robertson and Mullens, Melbourne

Price, J. 1798, quoted in 'The Koala' by E. Troughton, from A *Treasury of Australian Wildlife*, ed D.F. McMichael, Ure Smith, Sydney, p.55

Reed, A.W. 1965, *Myths and Legends of Australia*, A.H. and A.W. Reed, Sydney, Wellington, London

Reed, P.C., Lunney, D. and Walker, P. 1988, 'A 1986–87 Survey of the Koala, *Phascolarctos cinereus* (Goldfuss), in New South Wales and an Ecological Interpretation of its distribution', paper presented to the Third Symposium on the Biology of the Koala, Melbourne, February 1988

Robinson, G.A. 1844, in 'George Augustus Robinson's Journey into South-eastern Australia, 1844' ed G. Mackaness, *Australian Historical Monographs* Vol. 19, Review Publications, Dubbo, NSW

Robinson, A.C. 1978, 'The Koala in South Australia' in *The Koala, Proceedings of the Taronga Symposium*, ed T.J. Bergin, John Sands Pty Ltd, Artarmon, NSW

Serventy, V. 1988, 'Report of the 1949 Koala Survey in New South Wales' quoted in 'A 1986–87 Survey of the Koala, *Phascolarctos cinereus*, in New South Wales and an Ecological Interpretation of its Distribution' by P.C. Reed, D. Lunney and P. Walker, paper presented to the Third Symposium on the Biology of the Koala, Melbourne, February, 1988

Smith, M. 1980, 'Behaviour of the Koala, *Phascolarctos cinereus* (Goldfuss), in Captivity, *Australian Wildlife Research*, Nos. 6 and 7

Smith, M. 1987, 'Behaviour and Ecology' in *Koala, Australia's Endearing Marsupial*, ed by L. Cronin, Reed Books Pty Ltd, Frenchs Forest, NSW, pp.56–78

Stead, D.G. 1934, 'The Koala or Native Bear', *Australian Wildlife*, Vol. 1, pp.13–22

Storz, J. 1971, '*Chlamydia* and *Chlamydia*–induced Disease', Charles C. Thomas, Springfield, Illinois

Strahan, R. and Martin, R.W. 1982, 'The Koala: Little Fact, Much Emotion' in *Species at Risk: Research in Australia*, eds R.H. Grose and W.D.L. Ride, Australian Academy of Science, Canberra, pp. 147–55

Stivens, D. 1963, 'The Koala is Being Rescued', *Walkabout*, Vol. 29(5), pp.30–1

Strzelecki, P.E. 1840, quoted in *The Count: a Life of Sir Paul Edmund Strzelecki*, KCMG, *Explorer and Scientist*, William Heineman Ltd, Sydney

Troughton, E. Le G. 1935, '*Phascolarctos cinereus* Victor', *Australian Naturalist*, Vol. 9, p. 139

Troughton, E. Le G. 1941, *Furred Animals of Australia*, Angus and Roberston, Sydney

Troughton, E. Le G. 1967, 'The Koala' in A*Treasury of Australian Wildlife*, ed by D.F. McMichael, Ure Smith, Sydney, pp.45–60

Walsh, G.L. 1985, *Didane the Koala*, University of Queensland Press, St. Lucia, Queensland

Warneke, R.M. 1978, 'The Status of the Koala in Victoria', in *The Koala, Proceedings of the Taronga Symposium*, ed T.J. Bergin, John Sands Pty Ltd, Artarmon, NSW. pp.xxx

Weigler, B.J., Girjes, A.A., White, N.A., Kunst, N.D., Carrick, F.N. and Lavin, M.F. 1988, 'Aspects of the Epidemiology of *Chlamydia psittaci* Infection in a Population of Koalas (*Phascolarctos cinereus*) in South-eastern Queensland, *Journal of Wildlife Diseases*, Vol. 24, pp. 282–91

Wells, K.F., Wood, N.H. and Laut, P. 1984, 'Loss of Forests and Woodlands in Australia: a Summary by State Based on Rural Local Government Areas', CSIRO *Division of Land and Water Resources Technical Manual*, Vol. 84, No. 4, CSIRO, Melbourne

Wood-Jones, F. 1924, *The Mammals of South Australia Part II: the Bandicoots and the Herbivorous Marsupials* (the syndactylous Didelphia), Government Printer, Adelaide, pp.133–270

INDEX

V

W